THE EINSTEIN THEORY OF
RELATIVITY

THE EINSTEIN THEORY OF
RELATIVITY

A Trip to the Fourth Dimension

BY LILLIAN R. LIEBER

ILLUSTRATIONS BY HUGH GRAY LIEBER

EDITED AND WITH A NEW FOREWORD BY
DAVID DERBES AND ROBERT JANTZEN

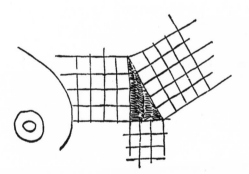

Paul Dry Books
PHILADELPHIA

2008

The editors would like to thank the keen-eyed Camillia Smith Barnes
for her careful help with this edition.

First Paul Dry Books Edition, 2008

Paul Dry Books, Inc.
Philadelphia, Pennsylvania
www.pauldrybooks.com

1 3 5 7 9 8 6 4 2
Printed in the United States of America

Library of Congress Cataloging-in-Publication Data

Lieber, Lillian R. (Lillian Rosanoff), 1886–1986.
  The Einstein theory of relativity : a trip to the fourth dimension /
by Lillian R. Lieber ; foreword by David Derbes and Robert Jantzen ;
illustrations by Hugh Gray Lieber. — 1st Paul Dry Books ed.
      p.   cm.
  Originally published: New York : Farrar & Rinehart, 1945. With new frwd.
  ISBN 978-1-58988-044-3 (alk. paper)
  1. Relativity (Physics) 2. Fourth dimension. 3. Einstein, Albert, 1879–1955.
I. Title.
  QC173.55.L517 2008
  530.11—dc22
                                                      2008020622

To

FRANKLIN DELANO ROOSEVELT

who saved the world from those forces
of evil which sought to destroy
Art and Science and the very
Dignity of Man.

CONTENTS

FOREWORD

Hundreds of books on Einstein and relativity have been written for lay readers, but the one you have in your hands is like no other. In all those other books, the authors strive to describe Einstein's work by analogy and metaphor, steering the reader as far as possible from the white water of mathematics and physics. Lillian and Hugh Lieber take an entirely different approach. Their little book was designed to teach—to almost anyone—the actual relativity of Albert Einstein, tensors and all. To follow all the mathematics in the second half of the book, you will need to know some differential calculus; but for most of the book, simple algebra and a little geometry will suffice. Even readers to whom mathematics is a foreign language can follow the arguments and see for themselves the breathtaking agreement between nature and Einstein's wonderful ideas.

Einstein's Theory of Relativity is actually two theories, or, more accurately, a theory that comes in two flavors, restricted and unrestricted. Einstein created the restricted theory in 1905 to resolve some troubling inconsistencies between the mechanical physics of Isaac Newton (1687) and the electric and magnetic physics of Michael Faraday (1844) and James Clerk Maxwell (1864). Einstein's theory ensured that measurements made by two observers in relative uniform (non-accelerated) motion would be consistent. At first, technology limited the experiments that could confirm or refute Einstein's theory, but over the past century, it has passed thousands of tests. More, Einstein's 1905 description of nature possesses a compelling simplicity and beauty absent from Newton's mechanics.

For ten years Einstein sought to generalize his principle to observers whose reference frames were moving with non-uniform velocity with respect to each other, or even to observers in any two frames related in any way whatsoever. Early on he felt that the extension to non-uniform motion might require, or otherwise include, the force of gravity. In 1915 he arrived at a generalized relativity, at its core a new theory of gravity. The original (restricted) Theory of Relativity then became known as *special relativity,* and the new (unrestricted) theory, Einstein's theory of gravity, was called *general relativity.*

Albert Einstein was not yet a household name when, in May of 1919, a solar eclipse visible in the southern hemisphere allowed for a sensitive test of general relativity. Astronomers traveled to the island of Principe off the coast of West Africa, and the town of Sobral in Brazil. Photographs taken during the eclipse were analyzed, and on November 6, 1919, during a special meeting of the Royal Society in London, the results were announced to the world. The data were fully consistent with Einstein's theory of gravity (but not with Newton's, which had predicted only half the observed value).

Until very recently, general relativity was taught only in postgraduate mathematics or physics courses, because the mathematical foundations of the theory were regarded as much too demanding for undergraduates. But the Liebers possessed an astounding, Promethean faith that a much larger audience could learn Einstein's theories—*the genuine article,* not watered-down explanations. They believed that Einstein's work, the deepest understanding of space and time yet conceived, belonged to all of us and should be made accessible to anyone who wanted to learn it. We share that belief. The first editions of this book were homemade by the Liebers (Hugh Lieber colored many of the illustrations by hand). After some years, a publisher took a chance, and kept the book in print for fifteen years. It has been out of print ever since, despite substantial efforts by the book's fans to get it republished. This new edition has made the dream of decades come true for us.

Many authors have described special relativity at about the same mathematical level as the first part of Professor Lieber's book, none with her economy or wit. The second part, describing general relativity, makes the book unique. The only other books providing such a close look at general relativity are textbooks; the authors of these books suppose their readers are already highly proficient with advanced mathematics. Professor Lieber assumes only geometry, hoping that you have a little calculus. If you do, and you're willing to grasp new tools, you will see *everything.* She shows you how to use these new tools (partial derivatives, determinants, tensor analysis), offers encouragement, and sympathetically warns you of approaching complications. She demonstrates the difficulties Einstein had to overcome, but she does not expect you to duplicate his labor.

The book's second part begins by justifying the use of non-Euclidean geometry. The tools and techniques for describing distances on a curved surface are introduced. The Riemann tensor is derived and Einstein's equations are stated. In the most mathematical part of the book (the reader is given fair warning), the Schwarzschild solution is derived, and from it is obtained a geodesic equation. This equation describes the motion of objects near a massive object (like our own sun). It provides the basis for three spectacular tests. Einstein's theory of gravity does not give the same answers as Newton's. Professor Lieber shows the reader how these theories differ in mathematical detail, philosophy, and predictions. She does not write down every step, hoping to spare readers a blizzard of symbols. Instead, she refers those seeking more details to a well-known textbook by the eminent English astrophysicist Sir Arthur Eddington. In two tests the math is elementary, and she shows exactly how numbers come out of Einstein's theory, to be compared with experimental data. One test involves gravity's unexpected influence on the rate of time, and the shift of atomic spectra towards the red end of the rainbow. (No such effect occurs in Newton's theory of gravity.) Professor Lieber shows that the ratio of a spectral line's frequency on earth to its frequency near an astronomical object of mass m and radius R is in the ratio $1 + (m/R):1$ (if the mass is measured in suitable units)—and the reader will have seen exactly how this comes about.

As a ninth grader in upstate New York, Robert Jantzen found this book in his small village public library, read it, and thought that when he got to college it would be fascinating to study the ideas on a deeper level. Even at Princeton, no undergraduate course was available. Robert joined a handful of classmates who convinced a visiting relativist, Remo Ruffini, to offer a seminar in general relativity. Robert made relativity his life's work, never forgetting it was this little book that started him on his career. David Derbes came to the book indirectly, from another work in the Lieber canon, and used it to teach a few of his high-school students an elective in relativity. Each was unaware of the other's fondness for the book, until David visited Robert's web page, read the tribute Robert paid to the Liebers' work, and learned that his old friend (and college classmate) was also interested in seeing the book republished.

Part of our job as editors has been to replace outdated references with more recent works. While deleting most, we have kept the references to classics but added newer (and more accessible) works in the footnotes and the bibliography. We have also provided notes to fill in nearly all the details Professor Lieber delegated to references in the works of Eddington and others. These are for readers who would like to work through everything but who lack easy access to technical libraries. It is perfectly possible to skim over much of the mathematical detail if you choose, but why not give the math a try? The poet Millay was perhaps overstating it to say that "Euclid alone has looked on Beauty bare," but we do agree with the great Russian theorist Lev Landau that "general relativity is probably the most beautiful of all physical theories."

Few people are as famous as Einstein, and few theories possess the almost magical powers of his relativity. We know of many scientists and mathematicians who as teenagers stumbled onto this little book, and were inspired by it. As a supplement to advanced high-school physics or math courses it could excite and inspire many more. Teachers of physics and calculus will enjoy working anew (or for the first time) to derive the famous predictions stunningly confirmed by the perihelion of Mercury, the spectral lines of Sirius, and the bending of starlight by the sun—thereby enriching their own understanding of these beautiful, deep ideas. It would be a wonderful thing if Einstein's relativity became part of the high-school curriculum.

We hope that many people will take up the challenge offered by the Liebers. Getting through their book is a little bit like running a marathon. Reading the book may likewise be demanding, but the rewards are also great.

DAVID DERBES
University of Chicago Laboratory Schools

ROBERT JANTZEN
Villanova University

March, 2008

PREFACE

In this book on the Einstein Theory of Relativity
the attempt is made
to introduce just enough
mathematics
to HELP
and NOT to HINDER
the lay reader;
"lay" can of course apply to
various domains of knowledge —
perhaps then we should say:
the layman in Relativity.

Many "popular" discussions of
Relativity,
without any mathematics at all,
have been written.
But we doubt whether
even the best of these
can possibly give to a novice
an adequate idea of
what it is all about.
What is very clear when expressed
in mathematical language
sounds "mystical" in
ordinary language.

On the other hand,
there are many discussions,
including Einstein's own papers,
which are accessible to the
experts only.

We believe that
there is a class of readers
who can get very little out of
either of these two kinds of
discussion —
readers who know enough about
mathematics
to follow a simple mathematical presentation
of a domain new to them,
built from the ground up,
with sufficient details to
bridge the gaps that exist
FOR THEM
in both
the popular and the expert
presentations.

This book is an attempt
to satisfy the needs of
this kind of reader.

NOTE

This is not intended to be
free verse.
Writing each phrase on a separate line
facilitates rapid reading,
and everyone
is in a hurry
nowadays.

ABOUT FOOTNOTES AND ENDNOTES

Asterisks, daggers, and similar symbols indicate the author's original footnotes. Where a new footnote has been added, or the original footnote changed significantly, the footnote is marked thus: #. Bracketed numbers indicate the editors' new endnotes on pp. 327–347.

Part I

THE SPECIAL THEORY

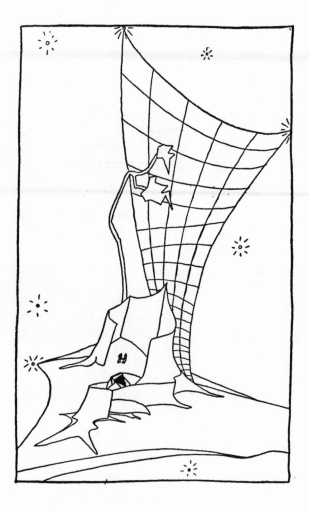

I. INTRODUCTION.

In order to appreciate
the fundamental importance
of Relativity,
it is necessary to know
how it arose.

Whenever a "revolution" takes place,
in any domain,
it is always preceded by
some maladjustment producing a tension,
which ultimately causes a break,
followed by a greater stability —
at least for the time being.

What was the maladjustment in Physics
in the latter part of the 19th century,
which led to the creation of
the "revolutionary" Relativity Theory?

Let us summarize it briefly:

It has been assumed that
all space is filled with ether,*
through which radio waves and light waves
are transmitted —
any modern child talks quite glibly

*This ether is of course NOT the chemical ether
which surgeons use!
It is not a liquid, solid, or gas,
it has never been seen by anybody,
its presence is only conjectured
because of the need for some medium
to transmit radio and light waves.

about "wavelengths"
in connection with the radio.

Now, if there is an ether,
does it surround the earth
and travel with it,
or does it remain stationary
while the earth travels through it?

Various known facts* indicate that
the ether does NOT travel with the earth.
If, then, the earth is moving THROUGH the ether,
then there must be an "ether wind,"
just as a person riding on a bicycle
through still air
feels an air wind blowing in his face.

And so an experiment was performed
by Michelson and Morley (see p. 8)
in 1887,
to detect this ether wind;
and much to the surprise of everyone,
no ether wind was observed.

This unexpected result was explained by
a Dutch physicist, Lorentz, in 1895,
in a way which will be described
in Chapter II.
The search for the ether wind
was then resumed
by means of other kinds of experiments.†

*See Sartori, *Understanding Relativity*,
  pp. 17-19 & pp. 111-116,
  in Further Reading.

†Ibid., pp. 26-51.

4

But, again and again,
to the consternation of the physicists,
no ether wind could be detected,
until it seemed that
nature was in a "conspiracy"
to prevent our finding this effect!

At this point
Einstein took up the problem,
and decided that
a natural "conspiracy"
must be a natural LAW operating.
And to answer the question
as to what is this law,
he proposed his Theory of Relativity,
published in two papers,
one in 1905 and the other in 1915.*

He first found it necessary to
re-examine the fundamental ideas
upon which classical physics was based,
and proposed certain vital changes in them.
He then made
A VERY LIMITED NUMBER OF
MOST REASONABLE ASSUMPTIONS
from which he deduced his theory.
So fruitful did his analysis prove to be
that by means of it he succeeded in:

(1) Clearing up the fundamental ideas.
(2) Explaining the Michelson-Morley experiment
    in a much more rational way than
    had previously been done.

*Both published by Dover in one volume,
 The Principle of Relativity,
 also including papers by Lorentz, Minkowski, and Weyl,
 with notes by Sommerfeld.
 Hereafter denoted POR. See [1] on p. 325.

(3) Doing away with
other outstanding difficulties
in physics.

(4) Deriving a
NEW LAW OF GRAVITATION
much more adequate than the
Newtonian one
(See Part II: The General Theory)
and which led to several
important predictions
which could be verified by experiment;
and which have been so verified
since then.

(5) Explaining
QUITE INCIDENTALLY
a famous discrepancy in astronomy
which had worried the astronomers
for many years
(This also is discussed in
The General Theory).

Thus, the Theory of Relativity had
a profound philosophical bearing
on ALL of physics,
as well as explaining
many SPECIFIC outstanding difficulties
that had seemed to be entirely
UNRELATED,
and of further increasing our knowledge
of the physical world
by suggesting a number of
NEW experiments which have led to
NEW discoveries.

No other physical theory
has been so powerful
though based on so FEW assumptions.
As we shall see.

II. THE MICHELSON-MORLEY EXPERIMENT.*

On page 4 we referred to
the problem that
Michelson and Morley set themselves.
Let us now see
what experiment they performed
and what was the startling result.

In order to get the idea of the experiment
very clearly in mind,
it will be helpful first
to consider the following simple problem,
which can be solved
by anyone who has studied
elementary algebra:

Imagine a river
in which there is a current flowing with
velocity v,
in the direction indicated by the arrow:

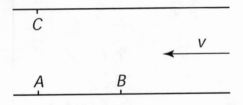

Now which would take longer—
for a man to swim
From A to B and back to A,

*Albert A. Michelson and Edward W. Morley,
"On the Relative Motion of the Earth and
the Luminiferous Ether,"
Am. Jour. Science XXXIV (1887): 333-345.
See Further Reading.

or
from A to C and back to A,
if the distances AB and AC are equal,
AB being parallel to the current,
and AC perpendicular to it?
Let the man's rate of swimming in still water
be represented by c;
then, when swimming against the current
from A to B,
his rate would be only $c - v$,
whereas,
when swimming with the current,
from B back to A,
his rate would, of course, be $c + v$.
Therefore the time required
to swim from A to B
would be $a/(c - v)$,
where a represents the distance AB;
and the time required
for the trip from B to A
would be $a/(c + v)$.
Consequently,
the time for the round trip would be

$$t_1 = \frac{a}{(c - v)} + \frac{a}{(c + v)}$$

or
$$t_1 = \frac{2ac}{(c^2 - v^2)}.$$

Now let us see
how long the round trip
from A to C and back to A
would take.
If he headed directly toward C,
the current would carry him downstream,
and he would land at some point
to the left of C in the figure on p. 8.
Therefore,
in order to arrive at C,

he should head for some point D
just far enough upstream
to counteract the effect of the current.

In other words,
if the water could be kept still
until he swam at his own rate c
from A to D,
and then the current
were suddenly allowed to operate,
carrying him at the rate v from D to C
(without his making any further effort),
then the effect would obviously be the same
as his going directly from A to C
with a velocity equal to $\sqrt{c^2 - v^2}$,
as is obvious from the right triangle:

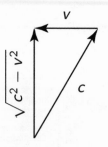

Consequently,
the time required
for the journey from A to C
would be $a/\sqrt{c^2 - v^2}$,
where a is the distance from A to C.
Similarly,
in going back from C to A,
it is easy to see that,

by the same method of reasoning,
the time would again be $a/\sqrt{c^2 - v^2}$.
Hence the time for the round trip
from A to C and back to A,
would be

$$t_2 = \frac{2a}{\sqrt{c^2 - v^2}}.$$

In order to compare t_1 and t_2 more easily,
let us write β for $c/\sqrt{c^2 - v^2}$. [2]
Then we get:

$$t_1 = 2a\beta^2/c$$
and $\quad t_2 = 2a\beta/c.$

Assuming that v is less than c,
and $c^2 - v^2$ being obviously less than c^2,
then $\sqrt{c^2 - v^2}$ is therefore less than c,
and consequently β is greater than 1
(since the denominator
is less than the numerator).
Therefore t_1 is greater than t_2,
that is,
IT TAKES LONGER TO
SWIM UPSTREAM AND BACK
THAN TO SWIM THE SAME DISTANCE
ACROSS-STREAM AND BACK.

But what has all this to do
with the Michelson-Morley experiment?
In that experiment,
a ray of light was sent from A to B:

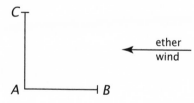

At B there was a mirror
which reflected the light back to A,
so that the ray of light
makes the round trip from A to B and back,
just as the swimmer did
in the problem described above.
Now, since the entire apparatus
shares the motion of the earth,
which is moving through space,
supposedly through a stationary ether,
thus creating an ether wind
in the opposite direction,
(namely, the direction indicated above),
this experiment seems entirely analogous
to the problem of the swimmer.
And, therefore, as before,

$$t_1 = 2a\beta^2/c \qquad (1)$$

and
$$t_2 = 2a\beta/c. \qquad (2)$$

Where c is now the velocity of light,
and t_2 is the time required for the light
to go from A to C and back to A
(being reflected from another mirror at C).
If, therefore,
t_1 and t_2 are found experimentally, —
then by dividing (1) by (2),
the value of β would be easily obtained.
And since $\beta = c/\sqrt{c^2 - v^2}$,
c being the known velocity of light,
the value of v could be calculated.
That is,
THE ABSOLUTE VELOCITY OF THE EARTH
would thus become known.

Such was the plan of the experiment.

Now what actually happened?

The experimental values of t_1 and t_2
were found to be the SAME,
instead of t_1 being greater than t_2!
Obviously this was a most disturbing result,
quite out of harmony
with the reasoning given above.
The Dutch physicist, Lorentz,
then suggested the following explanation
of Michelson's strange result:
Lorentz suggested that
matter, owing to its electrical structure,
SHRINKS WHEN IT IS MOVING,
and this contraction occurs
ONLY IN THE DIRECTION OF MOTION.*
The AMOUNT of shrinkage
he assumes to be in the ratio of $1/\beta$
(where β has the value $c/\sqrt{c^2 - v^2}$, as before).
Thus a sphere of one inch radius
becomes an ellipsoid when it is moving, [3]
with its shortest semi-axis
(now only $1/\beta$ inches long)

*The two papers by Lorentz on this subject
are included in POR, mentioned in
the footnote on page 5.
In the first of these papers
Lorentz mentions that the explanation proposed here
occurred also to Fitzgerald.
Hence it is often referred to as
the "Fitzgerald contraction" or
the "Lorentz contraction" or
the "Lorentz-Fitzgerald contraction."

in the direction of motion,
thus:

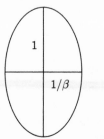

direction
of motion

Applying this idea
to the Michelson-Morley experiment,
the distance $AB\,(=a)$ on p. 8,
becomes a/β,
and t_1 becomes $2a\beta/c$,
instead of $2a\beta^2/c$,
so that now $t_1 = t_2$,
just as the experiment requires.

One might ask how it is
that Michelson did not
observe the shrinkage?
Why did not his measurements show
that AB was shorter than AC
(See the figure on p. 8)?
The obvious answer is that
the measuring rod itself contracts
when applied to AB,
so that one is not aware of the shrinkage.

To this explanation
of the Michelson-Morley experiment
the natural objection may be raised
that an explanation which is invented
for the express purpose

of smoothing out a certain difficulty,
and assumes a correction
of JUST the right amount,
is too artificial to be satisfying.
And Poincaré, the French mathematician,
raised this very natural objection.

Consequently,
Lorentz undertook to examine
his contraction hypothesis
in other connections,
to see whether it is in harmony also
with facts other than
the Michelson-Morley experiment.
He then published a second paper in 1904,
giving the result of this investigation.
To present this result in a clear form
let us first re-state the argument
as follows:

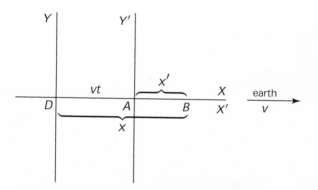

Consider a set of axes, X and Y,
supposed to be fixed in the stationary ether,
and another set X' and Y',
attached to the earth and moving with it,

with velocity v, as indicated above.
Let X' move along X,
and Y' move parallel to Y.

Now suppose an observer on the earth,
say Michelson,
is trying to measure
the time it takes a ray of light
to travel from A to B,
both A and B being fixed points on
the moving axis X'.
At the moment
when the ray of light starts at A
suppose that Y and Y' coincide,
and A coincides with D;
and while the light has been traveling to B
the axis Y' has moved the distance vt,
and B has reached the position
shown in the figure on p. 15,
t being the time it takes for this to happen.
Then, if $DB = x$ and $AB = x'$,
we have

$$x' = x - vt. \tag{3}$$

This is only another way
of expressing what was said on p. 9
where the time for
the first part of the journey
was said to be equal to $a/(c - v)$.*
And, as we saw there,
this way of thinking of the phenomenon
did NOT agree with the experimental facts.
Applying now the contraction hypothesis

---

*Since we are now designating a by x',
we have $x'/(c - v) = t$, or $x' = ct - vt$.
But the distance the light has traveled
is x,
and $x = ct$,
consequently $x' = x - vt$ is equivalent to $a/(c - v) = t$.

proposed by Lorentz,
x' should be divided by β,
so that equation (3) becomes

$$x'/\beta = x - vt$$

or
$$x' = \beta(x - vt). \tag{4}$$

Now when Lorentz examined other facts,
as stated on p. 15,
he found that equation (4)
was quite in harmony with all these facts,
but that he was now obliged
to introduce a further correction
expressed by the equation

$$t' = \beta(t - vx/c^2), \tag{5}$$

where $\beta, t, v, x,$ and c
have the same meaning as before —
But what is t'?!
Surely the time measurements
in the two systems are not different:
Whether the origin is at D or at A
should not affect the
TIME-READINGS.
In other words, as Lorentz saw it,
t' was a sort of "artificial" time
introduced only for mathematical reasons,
because it helped to give results
in harmony with the facts.
But obviously t' had for him
NO PHYSICAL MEANING.
As Jeans, the English physicist, puts it:
"If the observer could be persuaded
to measure time in this artificial way,
setting his clocks wrong to begin with
and then making them gain or lose permanently,
the effect of his supposed artificiality

17

would just counterbalance
the effects of his motion
through the ether"!*
Thus,
the equations finally proposed by Lorentz
are:

$$x' = \beta(x - vt)$$
$$y' = y$$
$$z' = z$$
$$t' = \beta(t - vx/c^2).$$

Note that
since the axes attached to the earth (p. 15)
are moving along the X-axis,
obviously the values of y and z
(z being the third dimension)
are the same as y' and z', respectively.

The equations just given
are known as
THE LORENTZ TRANSFORMATION,
since they show how to transform
a set of values of x, y, z, t
into a set x', y', z', t'
in a coordinate system
moving with constant velocity v,
along the X-axis,
with respect to the
unprimed coordinate system.
And, as we saw,
whereas the Lorentz transformation
really expressed the facts correctly,
it seemed to have
NO PHYSICAL MEANING,

*From Jeans' article, "Relativity," in the
14th edition of the *Encyclopedia Brittanica*.

and was merely
a set of empirical equations.

Let us now see what Einstein did.

III. RE-EXAMINATION OF THE
FUNDAMENTAL IDEAS.

As Einstein regarded the situation,
the negative result of
the Michelson-Morley experiment,
as well as of other experiments
which seemed to indicate a "conspiracy"
on the part of nature
against man's efforts to obtain
knowledge of the physical world (see p. 5),
these negative results,
according to Einstein,
did not merely demand
explanations of a certain number
of isolated difficulties,
but the situation was so serious
that a complete examination
of fundamental ideas
was necessary.
In other words,
he felt that there was something
fundamentally and radically wrong
in physics,
rather than a mere superficial difficulty.
And so he undertook to re-examine
such fundamental notions as
our ideas of
LENGTH and TIME and MASS.
His exceedingly reasonable examination

is most illuminating,
as we shall now see.

But first let us remind the reader
why length, time and mass
are fundamental.
Everyone knows that
VELOCITY depends upon
the distance (or LENGTH)
traversed in a given TIME,
hence the unit of velocity
DEPENDS UPON
the units of LENGTH and TIME.
Similarly,
since acceleration is
the change in velocity in a unit of time,
hence the unit of acceleration
DEPENDS UPON
the units of velocity and time,
and therefore ultimately upon
the units of LENGTH and TIME.
Further,
since force is measured
by the product of
mass and acceleration,
the unit of force
DEPENDS UPON
the units of mass and acceleration,
and hence ultimately upon
the units of
MASS, LENGTH and TIME.
And so on.
In other words,
all measurements in physics
depend primarily on
MASS, LENGTH and TIME.
That is why

the system of units ordinarily used
is called the "C.G.S." system,
where C stands for "centimeter"
(the unit of length),
G stands for "gram" (the unit of mass),
and S stands for "second" (the unit of time),
these being the fundamental units
from which all the others are derived.

Let us now return to
Einstein's re-examination of
these fundamental units.
Suppose that two observers
wish to compare their measurements of time.
If they are near each other
they can, of course, look at each other's watches
and compare them.
If they are far apart,
they can still compare each other's readings
BY MEANS OF SIGNALS,
say light signals or radio signals,
that is, any "electromagnetic wave"
which can travel through space.
Let us, therefore, imagine that
one observer, E, is on the earth,
and the other, S, on the sun;
and imagine that signals are sent
as follows:
By his own watch, S sends a message to E
which reads, "twelve o'clock";
E receives this message
say, eight minutes later;*

   *Since the sun is about 93,000,000 miles
    from the earth,
    and light travels about 186,000 miles per second,
    the time for a light (or radio) signal
    to travel from the sun to the earth
    is approximately eight minutes.

now, if his watch agrees with that of *S*,
it will read "12:08"
when the message arrives.
E then sends back to *S*
the message "12:08,"
and, of course,
S receives this message 8 minutes later,
namely, at 12:16.
Thus *S* will conclude,
from this series of signals,
that his watch and that of *E*
are in perfect agreement.

But let us know imagine
that the entire solar system
is moving through space,
so that both the sun and the earth
are moving in the direction
shown in the figure:

without any change in
the distance between them.
Now let the signals again be sent
as before:
S sends his message "12 o' clock,"
but since *E* is moving away from the message,
the latter will not reach *E* in 8 minutes,
but will take some longer time
to overtake *E*,
Say, 9 minutes.

If E's watch is in agreement with that of S,
it will read 12:09
when the message reaches him,
and E accordingly sends a return message,
reading "12:09."
Now S is traveling toward this message,
and it will therefore reach him
in LESS than 8 minutes,
say, in 7 minutes.
Thus S receives E's message
at 12:16,
just as before.
Now if S and E are both
UNAWARE of their motion
(and, indeed,
we are undoubtedly moving
in ways that we are entirely unaware of,
so that this assumption
is far from being an imaginary one).
S will not understand
why E's message reads
"12:09" instead of "12:08,"
and will therefore conclude
that E's watch
must be fast.
Of course, this is only
an apparent error in E's watch,
because, as we know,
it is really due to the motion,
and not at all
to any error in E's watch.
It must be noted, however,
that this omniscient "we"
who can see exactly
what is "really" going on in the universe,
does not exist,
and that all human observers

are really in the situation
in which *S* is,
namely,
that of not knowing
about the motion in question,
and therefore
being OBLIGED to conclude
that *E*'s watch is wrong!

And therefore,
S sends *E* the message
telling him that
if *E* sets his clock back one minute,
then their clocks will agree.

In the same way,
suppose that other observers,
A, *B*, *C*, etc.,
all of whom are at rest WITH RESPECT TO
S and *E*,
all set their clocks to agree with that of *S*,
by the same method of signals described above.
They would all say then
that all their clocks are in agreement.
Whether this is absolutely true or not,
they cannot tell (see above),
but that is the best they can do.

Now let us see what will happen
when these observers wish
to measure the length of something.
To measure the length of an object,
you can place it,
say, on a piece of paper,
put a mark on the paper at one end of the object,
and another mark at the other end,
then, with a ruler,
find out how many units of length there are

between the two marks.
This is quite simple provided that
the object you are measuring and the paper
are at rest (relative to you).
But suppose the object is
say, a fish swimming about in a tank?
To measure its length while it is in motion,
by placing two marks on the walls of the tank,
one at the head, and the other at the tail,
it would obviously be necessary
to make these two marks
SIMULTANEOUSLY —
for, otherwise,
if the mark *B* is made at a certain time,

then the fish allowed to swim
in the direction indicated by the arrow,
and then the mark at the head
is made at some later time,
when it has reached *C*,
then you would say that
the length of the fish
is the distance *BC*,
which would be a fish-story indeed!

Now suppose that our observers,
after their clocks are all in agreement (see p. 25),
undertake to measure
the length of a train

which is moving through their universe
with a uniform velocity.
They send out orders that
at 12 o'clock sharp,
whichever observer happens to be
at the place where
the front end of the train, A',
arrives at that moment,
to NOTE THE SPOT;
and some other observer,
who happens to be at the place where
the rear end of the train, B',
is at that same moment,
to put a mark at THAT spot.
Thus, after the train has gone,
they can, at their leisure,
measure the distance between the two marks,
this distance being equal to
the length of the train,
since the two marks were made
SIMULTANEOUSLY, namely at 12 o'clock,
their clocks being all
in perfect agreement with each other.

Let us now talk to the people on the train.
Suppose, first,
that they are unaware of their motion,
and that, according to them,
A, B, C, etc., are the ones who are moving, —
a perfectly reasonable assumption.
And suppose that there are two clocks on the train,
one at A', the other at B',
and that these clocks
have been set in agreement with each other
by the method of signals described above.
Obviously the observers A, B, C, etc.,
will NOT admit that the clocks at A' and B'

are in agreement with each other,
since they "know" that the train is in motion,
and therefore the method of signals
used on the moving train
has led to an erroneous setting
of the moving clocks (see p. 25).
Whereas the people on the train,
since they "know" that
A, B, C, etc., are the ones who are moving,
claim that it is the clocks
belonging to A, B, C, etc.,
which were set wrong.

What is the result of this
difference of opinion?
When the clocks of A and B, say,
both read 12 o'clock,
and at that instant A and B
each makes a mark at a certain spot,
then A and B claim, of course,
that these marks were made
simultaneously;
but the people on the train do not admit
that the clocks of A and B
have been properly set,
and they therefore claim that
the two marks were
NOT made SIMULTANEOUSLY,
and that, therefore,
the measurement of the LENGTH of the train
is NOT correct.
Thus,
when the people on the train
make the marks
simultaneously,
as judged by their own clocks,
the distance between the two marks

will NOT be the same as before.

Hence we see that
MOTION
prevents agreement in the
setting of clocks,
and, as a consequence of this,
prevents agreement in the
measurement of LENGTH!

Similarly,
as we shall see on p. 79,
motion also affects
the measurement of mass —
different observers obtaining
different results
when measuring the mass of the same object.

And since,
as we mentioned on p. 21,
all other physical measurements
depend upon
length, mass, and time,
it seems that
therefore there cannot be agreement
in any measurements made
by different observers
who are moving with different velocities!

Now, of course,
observers on the earth
partake of the various motions
to which the earth is subject —
the earth turns on its axis,
it goes around the sun,
and perhaps has other motions as well.
Hence it would seem that
observations made by people on the earth

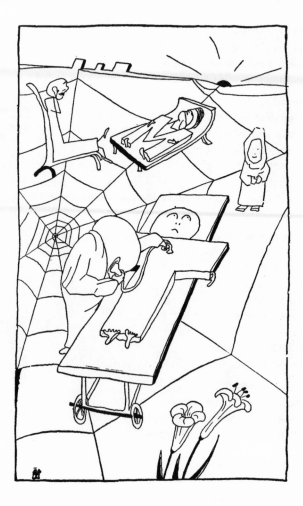

cannot agree with
those taken from
some other location in the universe,
and are therefore
not really correct
and consequently worthless!

Thus Einstein's careful and reasonable examination
led to the realization that
Physics was suffering from
no mere single ailment,
as evidenced by the
Michelson-Morley experiment alone,
but was sick from head to foot!

Did he find a remedy?

HE DID!

IV. THE REMEDY.

So far, then, we see that
THE OLD IDEAS REGARDING
THE MEASUREMENT OF
LENGTH, TIME AND MASS
involved an "idealistic" notion of
"absolute time"
which was supposed to be
the same for all observers,
and that
Einstein introduced
a more PRACTICAL notion of time
based on the actual way of
setting clocks by means of SIGNALS.
This led to the
DISCARDING of the idea that

the LENGTH of an object
is a fact about the object
and is independent of the person
who does the measuring,
since we have shown (Chapter III)
that the measurement of length
DEPENDS UPON
THE STATE OF MOTION OF THE MEASURER.

Thus two observers,
moving relative to each other
with uniform velocity,
DO NOT GET THE SAME VALUE
FOR THE LENGTH OF A GIVEN OBJECT.
Hence we may say that
LENGTH is NOT a FACT about an OBJECT,
but rather a
RELATIONSHIP between
the OBJECT and the OBSERVER.
And similarly for TIME and MASS (Ch. III).
In other words,
from this point of view
it is NOT CORRECT to say:

$$x' = x - vt$$

as Michelson did* (see p. 16, equation (3)),
since this equation implies that
the value of x'
is a perfectly definite quantity,

*We do not wish to imply that
Michelson made a crude error—
ANY CLASSICAL PHYSICIST
would have made the same statement,
for those were the prevailing ideas
thoroughly rooted in everybody's mind,
before Einstein pointed out
the considerations discussed in Ch. III.

namely,

THE length of the arm *AB* of the apparatus
in the Michelson-Morley experiment
(See the diagram on p. 15).
Nor is it correct to assume that

$$t' = t$$

(again as Michelson did)
for two different observers,
which would imply that
both observers agree in their
time measurements.

These ideas were contradicted by
Michelson's EXPERIMENTS,
which were so ingeniously devised
and so precisely performed.

And so Einstein said that
instead of starting with such ideas,
and basing our reasoning on them,
let us rather
START WITH THE EXPERIMENTAL DATA,
and see to what relationships
they will lead us,
relationships between
the length and time measurements
of different observers.
Now what experimental data
must we take into account here?
They are:

FACT (1): It is impossible
to measure the "ether wind,"
or, in other words,
it is impossible to detect our motion
relative to the ether.
This was clearly shown by the

Michelson-Morley experiment,
as well as by all other experiments
devised to
measure this motion (see p. 5).
Indeed, this is the great
"conspiracy"
that started all the trouble,
or, as Einstein prefers to see it,
and most reasonably so,
THIS IS A FACT.

FACT (2): The velocity of light is the same
no matter whether the source of light
is moving or stationary.
Let us examine this statement
more fully,
to see exactly what it means.

To do this,
it is necessary to remind the reader
of a few well-known facts:
Imagine that we have two trains,
one with a gun on the front end,
the other with a source of sound
on the front end,
say, a whistle.
Suppose that the velocity, u,
of a bullet shot from the gun,
happens to be the same as
the velocity of the sound.
Now suppose that both trains
are moving with the same velocity, v,
in the same direction.
The question is:
How does the velocity of a bullet
(fired from the MOVING train)
relative to the ground,
compare with

the velocity of the sound
that came from the whistle
on the other MOVING train,
relative to the medium, the air,
in which it is traveling?
Are they the same?

No!

The velocity of the bullet,
RELATIVE TO THE GROUND,
is $v + u$,
since the bullet is now propelled forward
not only with its own velocity, u,
given to it by the force of the gun,
but, in addition,
has an inertial velocity, v,
which it has acquired from
the motion of the train
and which is shared by
all objects on the train.

But in the case of the sound wave
(which is a series of pulsations,
alternate condensations and rarefactions of the air
in rapid succession),
the first condensation formed
in the neighborhood of the whistle,
travels out with the velocity u
relative to the medium,
regardless as to whether
the train is moving or not.
So that this condensation
has only its own velocity
and does NOT have the inertial velocity
due to the motion of the train,
the velocity of the sound
depending only upon the medium

(that is, whether it is air or water, etc.,
and whether it is hot or cold, etc.),
but not upon the motion of the source
from which the sound started.

The following diagram
shows the relative positions
after one second,
in both cases:

CASE I.
Both trains at rest.

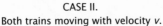

CASE II.
Both trains moving with velocity *v*.

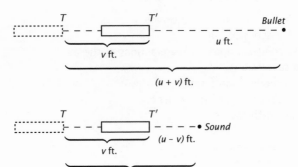

Thus, in Case II,
the bullet has moved
$u + v$ feet in one second

from the starting point,
whereas the sound has moved only u feet
from the starting point,
in that one second.
Thus we see that
the velocity of sound is u feet per second
relative to the starting point,
whether the source remains stationary
as in Case I,
or following the sound, as in Case II.

Expressing it algebraically,

$$x = ut$$

applies equally well for sound
in both Case I and Case II,
x being the distance
FROM THE STARTING POINT.
Indeed, this fact is true of ALL WAVE MOTION,
and one would therefore expect
that it would apply also to LIGHT.
As a matter of FACT,
it DOES,
and that is what is meant by
FACT (2) on p. 34.

Now, as a result of this,
it appears,
by referring again to the diagram on p. 36,
that
relative to the MOVING train (Case II)
we should then have,
for sound
$$x' = (u - v)t$$

x' being the distance
from T' to the point where
the sound has arrived after time t.

From which, by measuring x', u, and t,
we could then calculate v,
the velocity of the train.
And, similarly, for light
using the moving earth
instead of the moving train,
we should then have,
as a consequence of FACT (2) on p. 34,

$$x' = (c - v)t$$

where c is the velocity of light
(relative to a stationary observer
out in space)
and v is the velocity of the earth
relative to this stationary observer—
and hence
the ABSOLUTE velocity of the earth.

Then we should be able
to determine v.
But this contradicts FACT (1),
according to which
it is IMPOSSIBLE to determine v.

Thus it APPEARS that
FACT (2) requires
the velocity of light
RELATIVE TO THE MOVING EARTH
to be $c - v$ (see diagram on p. 36),
whereas FACT (1) requires it to be c.*

*FACT (1) may be re-stated as follows:
The velocity of light
RELATIVE TO A MOVING OBSERVER
(For example, an observer
on the moving earth)
must be c, and NOT $c - v$,
for otherwise,
he would be able to find v,
which is contrary to fact.

And so the two facts
contradict each other!
Or, stating it another way:

E E'

If, in one second,
the earth moves from E to E'
while a ray of light
goes from the earth to L,
then
FACT (1) requires that
$E'L$ be equal to c ($= 186,000$ miles)
while FACT (2) requires that
EL be equal to c!

Now it is needless to say that
FACTS CAN NOT CONTRADICT
EACH OTHER!

Let us therefore see how,
in the light of the discussion in Ch. III,
FACTS (1) and (2) can be shown to be
NOT contradictory.

V. THE SOLUTION OF THE DIFFICULTY.

We have thus seen that
according to the facts,
the velocity of light
IS ALWAYS THE SAME,

whether the source of light
is stationary or moving
(See FACT (2) on p. 34),
and whether the velocity of light
is measured
relative to the medium in which it travels,
or relative to a MOVING observer
(See p. 37).

Let us express these facts algebraically,
for two observers, K and K',
who are moving with uniform velocity
relative to each other,
thus:

$$K \text{ writes} \qquad x = ct, \tag{6}$$

$$\text{and } K' \text{ writes} \qquad x' = ct', \tag{7}$$

both using
THE SAME VALUE FOR
THE VELOCITY OF LIGHT,
namely, c,
and each using
his or her own measurements of
length, x and x',
and time, t and t', respectively.

It is assumed that
at the instant when
the rays of light start on their path,
K and K' are at the SAME place,
and the rays of light
radiate out from that place
in all directions.

Now according to equation (6),
K, who is unaware of his motion through the ether
(since he cannot measure it),
may claim that he is at rest,
and that in time, t,

K' must have moved to the right,
as shown in the figure below;
and that, in the meantime,
the light,
which travels out in all directions from K,
has reached all points at
the distance ct from K,
and hence
all points on the circumference
of the circle having the radius ct.

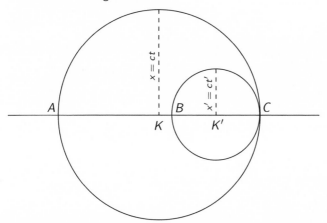

But K' claims that he is the one
who has remained stationary,
and that K, on the contrary,
has moved TO THE LEFT;
furthermore that the light travels out
from K' as a center,
instead of from K!
And this is what he means
when he says

$$x' = ct'.$$

How can they both be right?

41

We may be willing
not to take sides
in their controversy regarding the question as to
which one has moved —
K' to the right or K to the left —
because either leads to the same result.
But what about the circles?
They cannot possibly have both K and K'
as their centers!
One of them must be right and the other wrong.
This is another way of stating
the APPARENT CONTRADICTION BETWEEN
FACTS (1) and (2) (see p. 39).

Now, at last, we are ready
for the explanation.

Although K claims that
at the instant when
the light has reached the point C (p. 41),
it has also reached
the point A, on the other side,
still,
WE MUST REMEMBER THAT
when K says
two events happen simultaneously
(namely, the arrival of the light at C and A),
K' DOES NOT AGREE
THAT THEY ARE SIMULTANEOUS (see p. 28).
So that when
K' says that
the arrival of the light at C and B
(rather than at C and A),
ARE SIMULTANEOUS,
his statement
DOES NOT CONTRADICT THAT OF K,
since K and K'
DO NOT MEAN THE SAME THING

WHEN THEY SAY
"SIMULTANEOUS":
for
K's clocks at C and A
do not agree with K'''s clocks at C and A.
Thus when the light arrives at A,
the reading of K's clock there
is exactly the same as that of K's clock at C
(K having set all clocks in his system
by the method of signals described on p. 25),
while
K'''s clock at A,
when the light arrives there,
reads a LATER TIME than his clock at C
when the light arrived at C, —
so that K' maintains that
the light reaches A
LATER than it reaches C,
and NOT at the SAME instant,
as K claims.

Hence we see that
they are not really contradicting each other
but that they are merely using
two different systems of clocks,
such that
the clocks in each system
agree with each other all right,
but the clocks in the one system
have NOT been set
in agreement with the clocks
in the other system (see p. 28).

That is,
if we take into account
the inevitable necessity of
using signals
in order to set clocks which are

at a distance from each other,
and that the arrivals of the signals
at their destinations
are influenced by
our state of motion,
of which we are not aware (p. 24),
it becomes clear that
THERE IS NO REAL CONTRADICTION HERE,
but only a difference of description
due to INEVITABLE differences
in the setting of
various systems of clocks.

We now see
in a general qualitative way,
that the situation is
not at all mysterious or unreasonable,
as it seemed to be at first.
But we must now find out
whether these considerations,
when applied QUANTITATIVELY,
actually agree with the experimental facts.

And now a pleasant surprise awaits us.

VI. THE RESULT OF APPLYING
THE REMEDY.

In the last chapter we saw that
by starting with
two fundamental FACTS (p. 34),
we reached the conclusion
expressed in the equations

$$x = ct, \tag{6}$$
and
$$x' = ct' \tag{7}$$

which are graphically represented on p. 41,
and we realized that these equations
are NOT contradictory,
(as they appear to be at first),
if we remember that there is
a difference in the setting of the clocks
in the two different systems.

We shall derive, now, from (6) and (7),
relationships between the measurements
of the two observers, K and K'.
And all the mathematics we need for this
is a little simple algebra,
such as any high school student knows.

From (6) and (7) we get

$$x - ct = 0$$

and $$x' - ct' = 0.$$

Therefore

$$x' - ct' = \lambda(x - ct) \qquad (8)$$

where λ is a constant.
Similarly, in the opposite direction,

$$x' + ct' = \mu(x + ct) \qquad (9)$$

μ being another constant.
By adding and subtracting (8) and (9)

we get: $$x' = ax - bct \qquad (10)$$

and $$ct' = act - bx \qquad (11)$$

where $a = (\lambda + \mu)/2$ and $b = (\lambda - \mu)/2$.
Let us now find the values of a and b
in terms of v
(the relative velocity of K and K'),
and c, the velocity of light.

This is done in the following
ingenious manner:*
From (10)
when $x' = 0$
then $x = bct/a$;
but $x' = 0$ at the point K': (12)

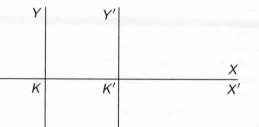

And x in this case is
the distance from K to K',
that is,
the distance traversed, in time t
by K' moving with velocity v
relative to K.
Therefore $x = vt$.
Comparing this with (12), we get

$$v = bc/a. \qquad (13)$$

Let us now consider the situation
from the points of view of K and K'.
Take K first:
From the time $t = 0$,
K gets $x' = ax$ (from (10)),

or $\qquad\qquad x = x'/a. \qquad (14)$

Hence K says that

*Einstein, *Relativity: The Special and General Theories*,
  Appendix I. See Further Reading.

46

to get the "true" value, x,

K' should divide his x' by a;

in particular,

if $x' = 1$,

K says that K''s unit of length

is only $1/a$ of a "true" unit.

But K',

at $t' = 0$, using (11)

says

$$bx = act \qquad (15)$$

and since from (10),

$$t = (ax - x')/bc,$$

(15) becomes

$$bx = ac(ax - x')/bc,$$

or $b^2 x = a^2 x - ax'$,

from which

$$x' = a\left(1 - b^2/a^2\right)x. \qquad (16)$$

And since $b/a = v/c$ from (13),

(16) becomes

$$x' = a\left(1 - v^2/c^2\right)x. \qquad (17)$$

In other words,

K' says:

In order to get the "true" value, x',

K should multiply her x by

$$a\left(1 - v^2/c^2\right).$$

In particular,

if $x = 1$,

then K' says that

K's unit is really $a\left(1 - v^2/c^2\right)$ units long.

Thus

each observer considers that
his or her own measurements
are the "true" ones,
and advises the other person
to make a "correction."
And indeed,
although the two observers, K and K',
may express this "correction"
in different forms,
still
the MAGNITUDE of the "correction"
recommended by each of them
MUST BE THE SAME,
since it is due in both cases
to the relative motion,
only that each observer attributes this motion
to the other person.

Hence, from (14) and (17) we may write*

$$1/a = a\left(1 - v^2/c^2\right).$$

Solving this equation for a, we get

$$a = c/\sqrt{c^2 - v^2}.$$

*Note that this equation is
NOT obtained by
ALGEBRAIC SUBSTITUTION
from (14) and (17),
but is obtained by considering that
the CORRECTIONS advised by K and K'
in (14) and (17), respectively,
must be equal in magnitude
as pointed out above.
Thus in (14) K says:
"You must multiply your measurement by $1/a$,"
whereas in (17) K' says:
"You must multiply your measurement by $a\left(1 - \frac{v^2}{c^2}\right)$";
and since these correction factors
must be equal
hence $1/a = a\left(1 - \frac{v^2}{c^2}\right)$.

Note that this value of a
is the same as that of β on p. 11.
Substituting in (10)
this value of a
and the value $bc = av$ from (13),
we get
$$x' = \beta x - \beta v t$$
or $\qquad\qquad x' = \beta (x - v t) \qquad\qquad\qquad$ (18)

which is the first of the set of equations
of the Lorentz transformation on page 19!

Furthermore,

from (18) and $\begin{cases} x = ct \\ x' = ct' \end{cases}$

we get
$$ct' = \beta (ct - v t)$$
or $\qquad\qquad t' = \beta (t - v t/c).$

Or, since $t = x/c$,
$$t' = \beta \left(t - v x/c^2 \right), \qquad\qquad\qquad (19)$$

which is another of the equations
of the Lorentz transformation!
That the remaining two equations
$y' = y$ and $z' = z$ also hold,
Einstein shows as follows:
Let K and K' each have a cylinder
of radius r, when at rest
relative to each other,
and whose axes coincide with the X (X') axis;
Now, unless $y' = y$ and $z' = z$,
K and K' would each claim that
his own cylinder is OUTSIDE the other person's!

We thus see that
the Lorentz transformation was derived
by Einstein
(quite independently of Lorentz),
NOT as a set of empirical equations

devoid of physical meaning,
but, on the contrary,
as a result of
a most rational change in
our ideas regarding the measurement of
the fundamental quantities
length and time.
And so, according to him,
the first of the equations of the
Lorentz transformation,
namely,

$$x' = \beta(x - vt)$$

is so written
NOT because of any real shrinkage,
as Lorentz supposed,
but merely an apparent shrinkage,*
due to the differences in
the measurements made by K and K' (see p. 45).
And Einstein writes

$$t' = \beta(t - vx/c^2)$$

NOT because it is just a mathematical trick
WITHOUT any MEANING (see p. 19)
but again because
it is the natural consequence of
the differences in the measurements
of the two observers.

And each observer may think
that he or she is right
and the other one is wrong,
and yet
each one,
by using his or her own measurements,
arrives at the same form

*This shrinkage, it will be remembered,
  occurs only in the direction of motion (see p. 13).

when he or she expresses a physical fact,
as, for example,
when K says $x = ct$
and K' says $x' = ct'$,
they are really agreeing as to
the LAW of the propagation of light.

And similarly,
if K writes any other law of nature,
and if we apply
the Lorentz transformation
to this law,
in order to see what form the law takes
when it is expressed in terms of
the measurements made by K',
we find that
the law is still the same,
although it is now expressed
in terms of the primed coordinate system.

Hence Einstein says that
although no one knows
what the "true" measurement should be,
yet,
each observer may use his or her own measurements
WITH EQUAL RIGHT AND EQUAL SUCCESS
in formulating
THE LAWS OF NATURE,
or,
in formulating the
INVARIANTS of the universe,
namely, the quantities which remain unchanged
in spite of the change in measurements
due to the relative motion of K and K'.

Thus, we can now appreciate
Einstein's Principle of Relativity:
"The laws by which
the states of physical systems

undergo change,
are not affected
whether these changes of state be referred
to the one or the other
of two systems of coordinates
in uniform translatory motion."

Perhaps some one will ask
"But is not the principle of relativity old,
and was it not known long before Einstein?"
Thus a person in a train
moving into a station
with uniform velocity
looks at another train which is at rest,
and imagines that the other train is moving
whereas his own is at rest.
And he cannot find out his mistake
by making observations within his train
since everything there
is just the same as it would be
if his train were really at rest.
Surely this fact,
and other similar ones,
must have been observered
long before Einstein?

In other words,
RELATIVE to an observer on the train
everything seems to proceed in the same way
whether his system (i.e., his train)
is at rest or in uniform motion,*
and he would therefore be unable

*Of course, if the motion is not uniform,
  but "jerky,"
  things on the train would jump around
  and the observer on the train
  would certainly know
  that his own train was not at rest.

53

to detect the motion.
Yes, this certainly was known
long before Einstein.
Let us see what connection it has
with the principle of relativity
as stated by him:

Referring to the diagram on p. 36
we see that
a bullet fired from a train
has the same velocity
RELATIVE TO THE TRAIN
whether the latter is moving or not,
and therefore an observer on the train
could not detect the motion of the train
by making measurements on
the motion of the bullet.
This kind of relativity principle
is the one involved
in the question on page 53,
and WAS known long before Einstein. [4]

Now Einstein
EXTENDED this principle
so that it would apply to
electromagnetic phenomena
(light or radio waves).

Thus,
according to this extension of
the principle of relativity,
an observer cannot detect
his motion through space
by making measurements on
the motion of ELECTROMAGNETIC WAVES.
But why should this extension
be such a great achievement —
why had it not been suggested before?

BECAUSE
it must be remembered that
according to fact (2) — see p. 39,

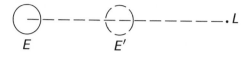

$$EL = c,$$
whereas,
the above-mentioned extension of
the principle of relativity
requires that $E'L$ should be equal to c
(compare the case of a bullet on p. 36).
In other words,
the extension of the principle of relativity
to electromagnetic phenomena
seems to contradict fact (2)
and therefore could not have been made
before it was shown that
fundamental measurements are merely "local"
and hence the contradiction was
only apparent,
as explained on p. 42;
so that the diagram shown above
must be interpreted
in the light of the discussion on p. 42.

Thus we see that
whereas the principle of relativity
as applied to MECHANICAL motion
(like that of the bullet)
was accepted long before Einstein,
the SEEMINGLY IMPOSSIBLE EXTENSION
of the principle
to electromagnetic phenomena
was accomplished by him.

This extension of the principle,
for the case in which
K and K' move relative to each other
with UNIFORM velocity,
and which has been discussed here,
is called
the SPECIAL theory of relativity.
We shall see later
how Einstein generalized this principle
STILL FURTHER,
to the case in which
K and K' move relative to each other
with an ACCELERATION,
that is, a CHANGING velocity.
And, by means of this generalization,
which he called
the GENERAL theory of relativity,
he derived
A NEW LAW OF GRAVITATION,
much more adequate even than
the Newtonian law,
and of which the latter
is a first approximation.

But before we can discuss this in detail
we must first see
how the ideas which we have
already presented
were put into a
remarkable mathematical form
by a mathematician named Minkowski.
This work
was essential to Einstein
in the further development of his ideas,
as we shall see.

VII. THE FOUR-DIMENSIONAL SPACE-TIME CONTINUUM.

We shall now see
how Minkowski* put Einstein's results [5]
in a remarkably neat mathematical form,
and how Einstein then utilized this
in the further application of his
Principle of Relativity,
which led to
The General Theory of Relativity,
resulting in a
NEW LAW OF GRAVITATION
and leading to further important consequences
and NEW discoveries.

It is now clear
from the Lorentz transformation (p. 19)
that
a length measurement, x',
in one coordinate system
depends upon BOTH x and t in another,
and that
t' also depends upon BOTH x and t.
Hence,
instead of regarding the universe
as being made up of
Space, on the one hand,
and Time, quite independent of Space,
there is a closer connection
between Space and Time
than we had realized.
In other words,

*H. Minkowski, "Space and Time," in POR.

that the universe is NOT a universe of points,
with time flowing along
irrespective of the points,
but rather,
this is
A UNIVERSE OF EVENTS, —
everything that happens,
happens at a certain place
AND at a certain time.

Thus, every event is characterized
by the PLACE and TIME of its occurrence.

Now,
since its place may be designated
by three numbers,
namely,
By the x, y, and z co-ordinates of the place
(using any convenient reference system),
and since the time of the event
needs only one number to characterize it,
we need in all
FOUR NUMBERS
TO CHARACTERIZE AN EVENT,
just as we need
three numbers to characterize
a point in space.

Thus we may say that
we live in a
four-dimensional world.
This does NOT mean
that we live in four-dimensional Space,
but is only another way of saying
that we live in
A WORLD OF EVENTS
rather than of POINTS only,
and it takes

FOUR numbers to designate
each significant element,
namely, each event.

Now if an event is designated
by the four numbers x, y, z, t,
in a given coordinate system,
the Lorentz transformation (p. 19)
shows how to find
the coordinates x', y', z', t',
of the same event,
in another coordinate system,
moving relative to the first
with uniform velocity.

In studying "graphs"
every high school freshman learns
how to represent a point
by two coordinates, x and y,
using the Cartesian system of coordinates,
that is,
two straight lines
perpendicular to each other.
Now, we may also use
another pair of perpendicular axes,
X' and Y' (in the figure on the next page),
having the same origin, 0, as before,
and designate the same point by x' and y'
in this new coordinate system.
When the high school student above-mentioned
studies analytical geometry,
he or she then learns how to find

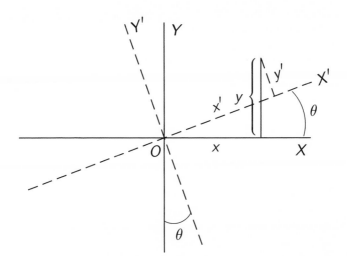

the relationship between
the primed coordinates
and the original ones,
and finds this to be expressed as follows:*

$$\left.\begin{array}{l} x = x' \cos \theta - y' \sin \theta \\ \text{and} \quad y = x' \sin \theta + y' \cos \theta \end{array}\right\} \quad (20)$$

where θ is the angle through which
the axes have been revolved,
as shown in the figure above.

The equations (20) remind one somewhat
of the Lorentz transformation (p. 19),
since the equations of

*See p. 310.

61

the Lorentz transformation
also show how to go
from one coordinate system to another.

Let us examine the similarity
between (20) and the
Lorentz transformation
a little more closely,
selecting from the
Lorentz transformation
only those equations involving x and t,
and disregarding those containing y and z,
since the latter remain unchanged
in going from one coordinate system to the other.
Thus we wish to compare (20) with:

$$\begin{cases} x' = \beta\,(x - vt) \\ t' = \beta\,(t - vx/c^2)\,. \end{cases}$$

Or, if, for simplicity, we take $c = 1$,
that is, taking
the distance traveled by light in one second,
as the unit of distance,
we may say that
we wish to compare (20) with

$$\left.\begin{array}{l} x' = \beta\,(x - vt) \\ t' = \beta\,(t - vx) \end{array}\right\}. \qquad (21)$$

Let us first solve (21) for x and t,
so as to get them more nearly
in the form of (20).
By ordinary algebraic operations,*

*And remembering that
we are taking $c = 1$,
and that therefore
$$\beta = \frac{1}{\sqrt{1-v^2}}$$

62

we get

$$x = \beta (x' + vt')$$
and
$$t = \beta (t' + vx')$$ \quad (22)

Before we go any further,
let us linger a moment
and consider equations (22):
Whereas (21) represents K speaking,
and saying to K':
"Now you must divide x' by β,
before you can get the relationship
between x and x' that you expect,
namely, equation (3) on p. 16;
in other words, your x' has shrunk
although you don't know it."

In (22),
it is K' speaking,
and he tells K the same thing,
namely that K must divide x by β,
to get the "true" x,
which is equal to $x' + vt'$.
Indeed,
this is quite in accord
with the discussion in Chapter VI,
in which it was shown that
each observer
gives the other one
precisely the same advice!
Note that the only difference
between (21) and (22) is that

$$+ v \text{ becomes } - v$$

in going from one to the other.

And this is again quite in accord
with our previous discussion —
since each observer
believes himself or herself to be at rest,
and the other person to be in motion,
only that one says:
"You have moved to the right" $(+v)$,
whereas the other says:
"You have moved to the left" $(-v)$.
Otherwise,
their claims are precisely identical;
and this is exactly what
equations (21) and (22) show so clearly.

Let us now return to the comparison
of (22) and (20):
Minkowski pointed out that
if, in (22),
t is replaced by $i\tau$ (where $i = \sqrt{-1}$),
and t' by $i\tau'$,
then (22) becomes: [6]

$$\begin{cases} x = \beta\,(x' + iv\tau') \\ i\tau = \beta\,(i\tau' + vx') \end{cases}$$

or

$$\begin{cases} x = \beta x' + i\beta v\tau' \\ i\tau = i\beta\tau' + \beta vx'. \end{cases}$$

Or (by multiplying the second equation by $-i$):

$$\begin{cases} x = \beta x' + i\beta v\tau' \\ \tau = \beta\tau' - i\beta vx'. \end{cases}$$

Finally,

substituting* $\cos\theta$ for β and $\sin\theta$ for $-i\beta v$
these equations become

$$\begin{cases} x = x'\cos\theta - \tau'\sin\theta \\ \tau = x'\sin\theta + \tau'\cos\theta \end{cases} \qquad (23)$$

EXACTLY like (20)!

In other words,
if K observes a certain event
and finds that
the four numbers she needs
to characterize it (see p. 58)
are x, y, z, τ,
and K', observing the SAME event,
finds that in his system
the four numbers
are x', y', z', τ',
then the form (23)
of the Lorentz transformation
shows that
to go from one observer's coordinate system
to the other
it is merely necessary
to rotate the first coordinate system
through an angle θ in the x, τ plane,
without changing the origin,

---

*Since β is greater than 1 (see p. 11)
θ must be an imaginary angle:
See p. 25 of *Non-Euclidean Geometry*,
another book by H. G. and L. R. Lieber.
Note that $\sin^2\theta + \cos^2\theta = 1$
holds for imaginary angles
as well as for real ones;
hence the above substitutions are legitimate,
thus $\beta^2 + (-i\beta v)^2 = \beta^2 - \beta^2 v^2 = \beta^2(1 - v^2) = 1$
since $\beta^2 = 1/(1 - v^2)$,
c being taken equal to 1 (see p. 62).

thus:

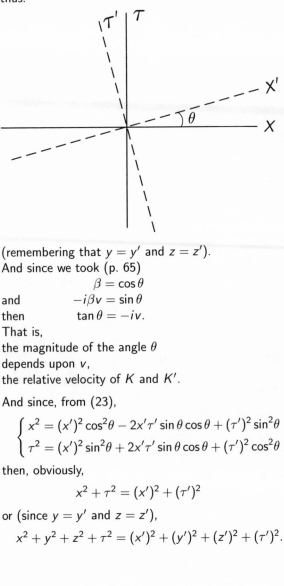

(remembering that $y = y'$ and $z = z'$).
And since we took (p. 65)

$$\beta = \cos \theta$$

and

$$-i\beta v = \sin \theta$$

then

$$\tan \theta = -iv.$$

That is,
the magnitude of the angle θ
depends upon v,
the relative velocity of K and K'.

And since, from (23),

$$\begin{cases} x^2 = (x')^2 \cos^2\theta - 2x'\tau' \sin\theta \cos\theta + (\tau')^2 \sin^2\theta \\ \tau^2 = (x')^2 \sin^2\theta + 2x'\tau' \sin\theta \cos\theta + (\tau')^2 \cos^2\theta \end{cases}$$

then, obviously,

$$x^2 + \tau^2 = (x')^2 + (\tau')^2$$

or (since $y = y'$ and $z = z'$),

$$x^2 + y^2 + z^2 + \tau^2 = (x')^2 + (y')^2 + (z')^2 + (\tau')^2.$$

Now, it will be remembered
from Euclidean plane geometry,

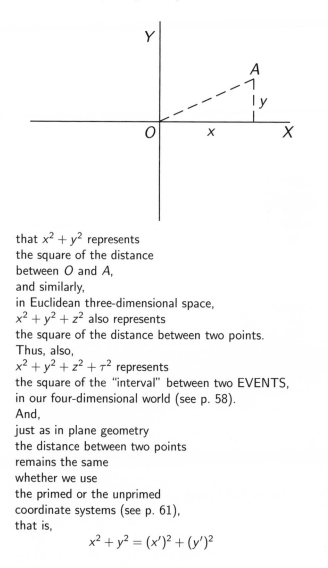

that $x^2 + y^2$ represents
the square of the distance
between O and A,
and similarly,
in Euclidean three-dimensional space,
$x^2 + y^2 + z^2$ also represents
the square of the distance between two points.
Thus, also,
$x^2 + y^2 + z^2 + \tau^2$ represents
the square of the "interval" between two EVENTS,
in our four-dimensional world (see p. 58).
And,
just as in plane geometry
the distance between two points
remains the same
whether we use
the primed or the unprimed
coordinate systems (see p. 61),
that is,

$$x^2 + y^2 = (x')^2 + (y')^2$$

(although x does NOT equal x',
and y does NOT equal y').
So, in three dimensions,

$$x^2 + y^2 + z^2 = (x')^2 + (y')^2 + (z')^2$$

and, similarly,
as we have seen on p. 66,
the "interval" between two events,
in our four-dimensional
space-time world of events,
remains the same,
no matter which of the two observers,
K or K',
measures it.

That is to say,
although K and K'
do not agree on some things,
as, for example,
their length and time measurements,
they DO agree on other things:

(1) The statement of their LAWS (see p. 51)
(2) The "interval" between events,
Etc.

In other words,
although length and time
are no longer INVARIANTS,
in the Einstein theory,
other quantities,
like the space-time interval between two events,
ARE invariants
in the theory.

These invariants are the quantities

which have the SAME value
for all observers,*
and may therefore be regarded
as the realities of the universe.

Thus, from this point of view,
NOT the things that we see or measure
are the realities,
since various observers
do not get the same measurements
of the same objects,
but rather
certain mathematical relationships
between the measurements
$\left(\text{Like } x^2 + y^2 + z^2 + \tau^2 \right)$
are the realities,
since they are the same
for all observers.*

We shall see,
in discussing
The General Theory of Relativity,
how fruitful
Minkowski's view-point of a
four-dimensional Space-Time World
proved to be.

VIII. SOME CONSEQUENCES OF THE THEORY OF RELATIVITY.

We have seen that
if two observers, K and K', move
relative to each other

*All observers moving relative to each other
with UNIFORM velocity (see p. 56).

with constant velocity,
their measurements of length and time
are different;
and, on page 29,
we promised also to show
that their measurements of mass are different.
In this chapter we shall discuss
mass measurements,
as well as other measurements which
depend upon
these fundamental ones.

We already know that if an object moves
in a direction parallel to
the relative motion of K and K',
then the Lorentz transformation
gives the relationship
between the length and time measurements
of K and K'.

We also know that
in a direction PERPENDICULAR to
the relative motion of K and K'
there is NO difference in the
LENGTH measurements (See footnote on p. 50),
and, in this case,
the relationship between the time measurements
may be found as follows:

For this PERPENDICULAR direction
Michelson argued that
the time would be

$$t_2 = 2a\beta/c \quad \text{(see p. 12)}.$$

Now this argument
is supposed to be from the point of view
of an observer who
DOES take the motion into account,

and hence already contains
the "correction" factor β;
hence,
replacing t_2 by t',
the expression $t' = 2a\beta/c$
represents the time
in the perpendicular direction
as K tells K' it SHOULD be written.
Whereas K, in her own system,
would, of course, write

$$t = 2a/c$$

for her "true" time, t.

Therefore

$$t' = \beta t$$

gives the relationship sought above,
from the point of view of K.

From this we see that
a body moving with velocity u
in this PERPENDICULAR direction,
will appear to K and K' to have
different velocities:
Thus,
since $u = d/t$ and $u' = d'/t'$
where d and d' represent
the distance traversed by the object
as measured by K and K', respectively;
and since $d = d'$
(there being NO difference in
LENGTH measurements in this direction —
see p. 70)
and $t' = \beta t$, as shown above,
then $u' = d/\beta t = (1/\beta)u$.

Similarly,
since $a = u/t$ and $a' = u'/t'$

where a and a' are the
accelerations of the body,
as measured by K and K', respectively,
we find that

$$a' = (1/\beta^2)a.$$

In like manner
we may find the relationships
between various quantities in the
primed and unprimed systems of co-ordinates,
provided they depend upon
length and time.

But, since there are THREE basic units in Physics
and since the Lorentz transformation
deals with only two of them, length and time,
the question now is
how to get the MASS into the game.
Einstein found that the best approach
to this difficult problem was via the
Conservation Laws of Classical Physics.
Then, just as the old concept of
the distance between two points
(three-dimensional)
was "stepped-up" to the new one of
the interval between two events
(four-dimensional), (see p. 67)
so also the Conservation Laws
will have to be "stepped up" into
FOUR-DIMENSIONAL SPACE-TIME.
And when this is done
an amazing vista will come into view!

CONSERVATION LAWS OF
CLASSICAL PHYSICS:
(1) Conservation of Mass: this means that
    no mass can be created or destroyed,

but only transformed from one kind to another.
Thus, when a piece of wood is burned,
its mass is not destroyed, for
if one weighs all the substances into which
it is transformed, together with the ash
that remains, this total weight is the
same as the weight of the original wood.
We express this mathematically thus: $\Delta\Sigma m = 0$
where Σ stands for the SUM, so that Σm
is the TOTAL mass, and Δ, as usual,
stands for the "change," so that
$\Delta\Sigma m = 0$ says that the change in
total mass is zero, which is the
Mass Conservation Law in very
convenient, brief, exact form!

(2) Conservation of Momentum: this says that
if there is an exchange of momentum
(the product of mass and velocity, mv)
between bodies, say, by collision, the
TOTAL momentum BEFORE collision
is the SAME
as the TOTAL after collision: $\Delta\Sigma mv = 0$.

(3) Conservation of Energy: which means that
Energy cannot be created or destroyed, but
only transformed from one kind to another.
Thus, in a motor, electrical energy is
converted to mechanical energy, whereas
in a dynamo the reverse change takes place.
And if, in both cases, we take into account
the part of the energy which is
transformed into heat energy, by friction,
then the TOTAL energy
BEFORE and AFTER the transformation
is the SAME, thus: $\Delta\Sigma E = 0$.
Now, a moving body has

KINETIC energy, expressible thus: $\frac{1}{2}mv^2$.
When two moving, ELASTIC bodies collide,
there is no loss in kinetic energy of
the whole system, so that then we have
Conservation of Kinetic Energy: $\Delta\Sigma\frac{1}{2}mv^2 = 0$
(a special case of the more general Law);
whereas, for inelastic collision, where
some of the kinetic energy is changed into
other forms, say heat, then $\Delta\Sigma\frac{1}{2}mv^2 \neq 0$.

Are you wondering what is the use of all this?
Well, by means of these Laws, the most
PRACTICAL problems can be solved,*
hence we must know what happens to them
in Relativity Physics!
You will see that they will lead to:

(a) NEW Conservation Laws for
Momentum and Energy, which are
INVARIANT under
the Lorentz transformation,
and which reduce, for small v, ‡
to the corresponding Classical Laws
(which shows why those Laws worked
so well for so long!)

(b) the IDENTIFICATION of
MASS and ENERGY!
Hence mass CAN be destroyed as such
and actually converted into energy!
Witness the ATOMIC BOMB (see p. 318).

* See any introductory physics textbook,
for example, Holton & Brush,
Physics, the Human Adventure, pp. 209–229
in Further Reading.

‡ Remembering that the "correction" factor, β, is equal to
$c/\sqrt{c^2 - v^2}$, you see that, when v is small relative to
the velocity of light, c, thus making v^2 negligible, then
$\beta = 1$ and hence no "correction" is necessary.

Thus the Classical Mass Conservation Law
was only an approximation and becomes
merged into the Conservation of Energy Law!

Even without following the mathematics of
the next few pages, you can already
appreciate the revolutionary IMPORTANCE of
these results, and become imbued with
the greatest respect for the human MIND
which can create all this and
PREDICT happenings previously unknown!
Here is MAGIC for you!

Some readers may be able to understand
the following "stepping up" process now,
others may prefer to come back to it
after reading Part II of this book:

The components of the velocity vector
in Classical Physics, are:

$$dx/dt, \ dy/dt, \ dz/dt.$$

And, if we replace x, y, z by x_1, x_2, x_3,
these become, in modern compact notation:

$$dx_i/dt \qquad (i = 1, 2, 3).$$

Similarly, the momentum components are :

$$m \cdot dx_i/dt \qquad (i = 1, 2, 3)$$

so that, for n objects,
the Classical Momentum Conservation Law is:

$$\Delta \left\{ \sum_n m \cdot dx_i/dt \right\} = 0 \qquad (i = 1, 2, 3) \qquad (24)$$

But (24) is NOT an invariant under
the Lorentz transformation;

the corresponding vector which IS
so invariant is:

$$\Delta \left\{ \sum_n m \cdot dx_i/ds \right\} = 0 \qquad (i = 1, 2, 3, 4) \qquad (25)$$

where s is the interval between two events,
and it can be easily shown* that $ds = dt/\beta$,
ds being, as you know, itself invariant
under the Lorentz transformation.
Thus, in going from 3-dimensional space
and 1-dimensional absolute time
(i.e. from Classical Physics)
to 4-dimensional SPACE-TIME,
we must use s for the independent variable
instead of t.
Now let us examine (25) which is so easily
obtained from (24) when we learn to speak the
NEW LANGUAGE OF SPACE-TIME!
Consider first only the first 3 components of (25):

Then $\qquad \Delta \left\{ \sum_n m \cdot dx_i/ds \right\} = 0 \qquad (i = 1, 2, 3) \qquad (26)$

is the NEW Momentum Conservation Law,
since, for large v, it holds whereas (24) does NOT;
and, for small v, which makes $\beta = 1$ and $ds = dt$,
(26) BECOMES (24), as it should!
And now, taking the FOURTH component of (25),
namely, $m \cdot dx_4/ds$ or $mc \cdot dt/ds$ (see p. 233)
and substituting dt/β for ds,
we get $mc\beta$ which is $mc \cdot c/\sqrt{c^2 - v^2}$ or

$$mc/\sqrt{1 - v^2/c^2} = mc \left(1 - v^2/c^2\right)^{-\frac{1}{2}}. \qquad (27)$$

Expanding, by the binomial theorem, [7] we get

$$mc \left(1 + \tfrac{1}{2} \cdot v^2/c^2 + \tfrac{3}{8} \cdot v^4/c^4 + \dots \right),$$

*Since $ds^2 = c^2dt^2 - (dx^2 + dy^2 + dz^2)$ (see p. 233),
dividing by dt^2 and taking $c = 1$, we get
$(ds/dt)^2 = 1 - v^2$ and $ds/dt = \sqrt{1 - v^2} = 1/\beta$.

which, for small v (neglecting terms after v^2),
becomes, approximately,

$$mc\left(1 + \tfrac{1}{2}v^2/c^2\right). \tag{28}$$

And, multiplying by c, we get $mc^2 + \tfrac{1}{2}mv^2$.
Hence, approximately,

$$\Delta\left\{\sum_n\left(mc^2 + \tfrac{1}{2}mv^2\right)\right\} = 0. \tag{29}$$

Now, if m is constant, as for elastic collision,
then $\Delta\Sigma mc^2 = 0$ and therefore also $\Delta\Sigma(\tfrac{1}{2}mv^2) = 0$
which is the Classical Law of the
Conservation of Kinetic Energy for
elastic collision (see p. 74);
thus (29) reduces to this Classical Law
for small v, as it should!
Furthermore, we can also see from (29) that
for INELASTIC collision, for which

$$\Delta\left\{\sum_n\tfrac{1}{2}mv^2\right\} \neq 0 \text{ (see p. 74)}$$

hence also $\Delta\Sigma mc^2 \neq 0$ or
c being a constant, $c^2\Delta\Sigma m \neq 0$
which says that, for inelastic collision,
even when v is small,
any loss in kinetic energy is compensated for
by an increase in mass (albeit small)
by a new and startling consequence for
CLASSICAL Physics itself!
Thus, from this NEW viewpoint we realize that
even in Classical Physics
the Mass of a body is NOT a constant but
varies with the change in its energy
(the amount of change in mass being
too small to be directly observed)!
Taking now (27) instead of (28), we shall
not be limited to small v;

and, multiplying by c as before,

we get $\quad \Delta \left\{ \sum\limits_{n} mc^2\beta \right\} = 0$ for the

NEW Conservation Law of Energy,
which, together with (25), is invariant under the
Lorentz transformation, and which,
as we saw above, reduces to
the corresponding Classical Law, for small v.
Thus the NEW expression for the ENERGY
of a body is: $E = mc^2\beta$, which,

for $v = 0$, gives $E_o = mc^2$, $\qquad\qquad\qquad$ (30)

showing that
ENERGY and MASS are
one and the same entity
instead of being distinct, as previously thought!
Futhermore,
even a SMALL MASS, m,
is equivalent to a LARGE amount of ENERGY,
since the multiplying factor is c^2,
the square of the enormous velocity of light!
Thus even an atom is equivalent to
a tremendous amount of energy.
Indeed, when a method was found (see p. 318)
of splitting an atom into two parts
and since the sum of these two masses is
less than the mass of the original atom,
you can see from (30) that
this loss in mass must yield
a terrific amount of energy
(even though this process does not transform
the entire mass of the original atom into energy).
Hence the ATOMIC BOMB! (p. 318)
Although this terrible gadget has
stunned us all into the realization
of the dangers of Science,
let us not forget that

the POWER behind it
is the human MIND itself.
Let us therefore pursue our examination of
the consequences of Relativity,
the products of this REAL POWER!

In 1901 (before Relativity),
Kaufmann [8], experimenting with
fast moving electrons,
found that
the apparent mass of a moving electron
is greater than that of one at rest —
a result which seemed
very strange at the time!
Now, however, with the aid of (26)
we can see
that his result is perfectly intelligible,
and indeed accounts for it quantitatively!
Thus the use of ds instead of dt,
(where $ds = dt/\beta$) brings in
the necessary correction factor, β, for large v,
not via the mass but is inherent in our
NEW RELATIVITY LANGUAGE,
in which dx_i/ds replaces the idea of
velocity, dx_i/dt, and makes it
unnecessary and undesirable to think in terms of
mass depending upon velocity.
Many writers on Relativity replace
ds by dt/β in (26) and write it:

$$\Delta \left\{ \sum_n m\beta \cdot dx_i/dt \right\} = 0, \text{ putting the}$$

correction on the m.
Though this of course gives

the same numerical result,
it is a concession to
CLASSICAL LANGUAGE,
and Einstein himself does not like this.
He rightly prefers that since we are
learning a NEW language (Relativity)
we should think directly in that language
and not keep translating each term
into our old CLASSICAL LANGUAGE
before we "feel" its meaning.
We must learn to "feel" modern and talk modern.

Let us next examine
another consequence of
the Theory of Relativity:

When radio signals are transmitted
as electromagnetic waves,
an observer K may measure
the electric and magnetic fields
at any point of the wave
at a given instant.
The relationship between
these electric and magnetic fields
is expressed mathematically
by the well-known Maxwell equations
(see page 311).

Now, if another observer, K',
moving relative to K
with uniform velocity,
makes his own measurements
on the same phenomenon,
and, according to
the Principle of Relativity,
uses the same Maxwell equations
in his primed system,

it is quite easy to show* that
the electric field
is NOT an INVARIANT
for the two observers;
and similarly
the magnetic field is also
NOT AN INVARIANT
although the relationship between
the electric and magnetic fields
expressed in the
MAXWELL EQUATIONS
has the same form for
both observers;
just as, on p. 68,
though x does NOT equal x'
and y does NOT equal y'
still the formula for
the square of the distance between two points
has the same form
in both systems of coordinates.

Thus we have seen that
the SPECIAL Theory of Relativity,
which is the subject of Part I (see p. 56),
has accomplished the following:

(1) It revised the fundamental physical concepts.

(2) By the addition of
    ONLY ONE NEW POSTULATE,
    namely,
    the extension of
    the principle of relativity

*See Einstein, "On the Electrodynamics of Moving Bodies,"
pp. 52–53, in POR.

to ELECTROMAGNETIC phenomena*
(which extension was made possible
by the above-mentioned revision
of fundamental units — see p. 55),
it explained many
ISOLATED experimental results
which baffled the
pre-Einsteinian physicists:
As, for example,
the Michelson-Morley experiment,
Kaufmann's experiments (p. 79),
and many others (p. 6).

(3) It led to the merging into
ONE LAW
of the two, formerly isolated, principles
of the Conservation of Mass and
the Conservation of Energy.

In Part II
we shall see also how
the SPECIAL Theory served as a
starting point for
the GENERAL THEORY,

*The reader may ask:
"Why call this a postulate?
Is it not based on facts?"
The answer of course is that
a scientific postulate must be
BASED on facts,
but it must go further than the known facts
and hold also for
facts that are still TO BE discovered.
So that it is really only an ASSUMPTION
(a most reasonable one, to be sure,
since it agrees with facts now known),
which becomes strengthened in the course of time
if it continues to agree with NEW facts
as they are discovered.

which, again,
by means of only
ONE other assumption,
led to FURTHER NEW IMPORTANT RESULTS,
results which make the theory
the widest in scope
of any physical theory.

IX. A POINT OF LOGIC AND A SUMMARY.

It is interesting here
to call attention to a logical point
which is made very clear
by the Special Theory of Relativity.
In order to do this effectively
let us first list and number
certain statements, both old and new,
to which we shall then refer by NUMBER:

(1) It is impossible for an observer
to detect his motion through space (p. 33).

(2) The velocity of light is
independent of the motion of the source (p. 34).

(3) The old PRE-EINSTEINIAN postulate
that time and length measurements
are absolute,
that is,
are the same for all observers.

(4) Einstein's replacement of this postulate
by the operational fact (see p. 31)
that
time and length measurements

are NOT absolute,
but relative to each observer.

(5) Einstein's Principle of Relativity (p. 52).

We have seen that
(1) and (2)
are contradictory IF (3) is retained
but are NOT contradictory IF
(3) is replaced by (4).   (Ch. V.)
Hence
it may NOT be true to say that
two statements MUST be
EITHER contradictory or NOT contradictory,
without specifying the ENVIRONMENT —
Thus,
in the presence of (3),
(1) and (2) ARE contradictory,
whereas,
in the presence of (4),
the very same statements (1) and (2)
are NOT contradictory.*
We may now briefly summarize
the Special Theory of Relativity:
(1), (2) and (4)
are the fundamental ideas in it,
and,
since (1) and (4) are embodied in (5),
then (2) and (5) constitute
the BASIS of the theory.

Einstein gives these two
as POSTULATES

*Similarly
  whether two statements are
  EQUIVALENT or not
  may also depend upon the environment
  (see p. 30 of *Non-Euclidean Geometry*
  by H. G. and L. R. Lieber).

from which he then deduces
the Lorentz transformation (p. 49)
which gives the relationship
between the length and time measurements*
of two observers moving relative to each other
with uniform velocity,
and which shows that
there is an intimate connection
between space and time.

This connection was then
EMPHASIZED by Minkowski,
who showed that
the Lorentz transformation may be regarded
as a rotation in the x, τ plane
from one set of rectangular axes to another
in a four-dimensional space-time continuum
(see Chapter VII).

*For the relationships between
other measurements,
see Chapter VIII.

THE MORAL.

1. Local, "provincial" measurements
   are not universal,
   although they may be used
   to obtain universal realities
   if compared with other systems of
   local measurements taken from
   a different viewpoint.
   By examining certain
   RELATIONSHIPS BETWEEN
   LOCAL MEASUREMENTS,
   and finding those relationships which
   remain unchanged in going from
   one local system to another,
   one may arrive at
   the INVARIANTS of our universe.

2. By emphasizing the fact that
   absolute space and time
   are pure mental fictions,
   and that the only PRACTICAL notions of time
   that man can have
   are obtainable only by some method of signals,
   the Einstein Theory shows that
   "Idealism" alone,
   that is, "a priori" thinking alone,
   cannot serve for exploring the universe.
   On the other hand,
   since actual measurements
   are local and not universal,

and that only certain
THEORETICAL RELATIONSHIPS
are universal,
the Einstein Theory shows also that
practical measurement alone
is also not sufficient
for exploring the universe.
In short,
a judicious combination
of THEORY and PRACTICE,
EACH GUIDING the other —
a "dialectical materialism" —
is our most effective weapon.

Part II

THE GENERAL THEORY

A GUIDE FOR THE READER.

I. The first three chapters of Part II give
the meaning of the term
"General Relativity,"
what it undertakes to do,
and what are its basic ideas.
These are easy reading and important.

II. Chapters XIII, XIV, and XV introduce
the fundamental mathematical ideas
which will be needed —
also easy reading and important.

III. Chapters XVI to XXII build up
the actual
streamlined mathematical machinery —
not difficult, but require
the kind of
care and patience and work
that go with learning to
run any NEW machine.
The amazing POWER of this new
TENSOR CALCULUS,
and the EASE with which it is operated,
are a genuine delight!

IV. Chapters XXIII to XXVIII show how
this machine is used to derive the
NEW LAW OF GRAVITATION.
This law,
though at first complicated

behind its seeming simplicity,
is then
REALLY SIMPLIFIED.

V. Chapters XXIX to XXXIV constitute
THE PROOF OF THE PUDDING! —
easy reading again —
and show
what the machine has accomplished.

Then there are
a SUMMARY
and
THE MORAL.

X. INTRODUCTION.

In Part I,
on the SPECIAL Theory,
it was shown that
two observers who
are moving relatively to each other
with UNIFORM velocity
can formulate
the laws of the universe
"WITH EQUAL RIGHT AND
EQUAL SUCCESS,"
even though
their points of view
are different,
and their actual measurements
do not agree.

The things that appear alike
to them both
are the "FACTS" of the universe,
the INVARIANTS.
The mathematical relationships
which both agree on
are the "LAWS" of the universe.
Since man does not know
the "true laws of God,"
why should any one human viewpoint
be singled out
as more correct than any other?
And therefore
it seems most fitting
to call THOSE relationships

"THE laws,"
which are VALID from
DIFFERENT viewpoints,
taking into consideration
all experimental data
known up to the present time.

Now, it must be emphasized
that in the Special Theory,
only that change of viewpoint
was considered
which was due to
the relative UNIFORM velocity
of the different observers.
This was accomplished by
Einstein
in his first paper*
published in 1905.
Subsequently, in 1916,
he published a second paper†
in which
he GENERALIZED the idea
to include observers
moving relatively to each other
with a CHANGING velocity
(that is, with an ACCELERATION),
and that is why it is called
"the GENERAL Theory of Relativity."

It was shown in Part I
that
to make possible
even the SPECIAL case considered there
was not an easy task,

---

* "On the Electrodynamics of Moving Bodies," in POR.
† "The Foundation of the General Theory of Relativity," in POR.

for it required
a fundamental change in Physics
to remove the
APPARENT CONTRADICTION
between certain
EXPERIMENTAL FACTS!
Namely,
the change from the OLD idea
that TIME is absolute
(that is,
that it is the same for all observers)
to the NEW idea that
time is measured
RELATIVELY to an observer,
just as the ordinary
space coordinates, x, y, z,
are measured relatively to
a particular set of axes.
This SINGLE new idea
was SUFFICIENT
to accomplish the task
undertaken in
the Special Theory.

We shall now see that
again
by the addition of
ONLY ONE more idea,
called
"THE PRINCIPLE OF EQUIVALENCE,"
Einstein made possible
the GENERAL Theory.

Perhaps the reader may ask
why the emphasis on the fact that
ONLY ONE new idea
was added?
Are not ideas good things?

And is it not desirable
to have as many of them as possible?
To which the answer is that
the adequateness
of a new scientific theory
is judged

(a) By its correctness, of course,
    and

(b) By its SIMPLICITY.

No doubt everyone appreciates
the need for correctness,
but perhaps
the lay reader may not realize
the great importance of
SIMPLICITY!

"But," he will say,
"surely the Einstein Theory
is anything but simple!
Has it not the reputation
of being unintelligible
to all but a very few experts?"

Of course,
"SIMPLE" does not necessarily mean
"simple to everyone," *
but only in the sense that

  *Indeed, it can even be simple to
  everyone WHO
  will take the trouble to learn some
  mathematics.
  Though this mathematics
  was DEVELOPED by experts,
  it can be UNDERSTOOD by
  any earnest student.
  Perhaps even the lay reader
  will appreciate this
  after reading this little book.

if all known physical facts
are taken into consideration,
the Einstein Theory accounts for
a large number of these facts
in the SIMPLEST known way.

Let us now see
what is meant by
"The Principle of Equivalence,"
and what it accomplishes.

It is impossible to refrain
from the temptation
to brag about it a bit
in anticipation!
And to say that
by making the General Theory possible,
Einstein derived
A NEW LAW OF GRAVITATION
which is even more adequate than
the Newtonian one,
since it explains,
QUITE INCIDENTALLY,
experimental facts
which were left unexplained
by the older theory,
and which had troubled
the astronomers
for a long time.

And, furthermore,
the General Theory
PREDICTED NEW FACTS,
which have since been verified —
this is of course
the supreme test of any theory.

But let us get to work
to show all this.

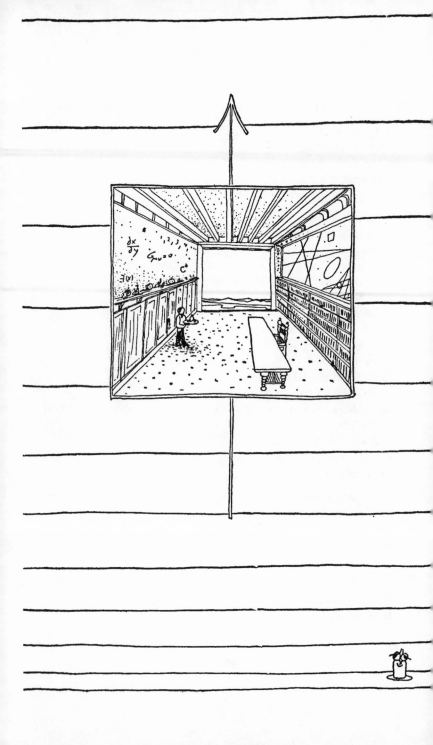

XI. THE PRINCIPLE OF EQUIVALENCE.

Consider the following situation:

Suppose that a man, Mr. K,
lives in a spacious box,
away from the earth
and from all other bodies,
so that there is no force of gravity
there.
And suppose that
the box and all its contents
are moving (in the direction
indicated in the drawing on p. 100)
with a changing velocity,
increasing 32 ft. per second
every second.
Now Mr. K,
who cannot look outside of the box,
does not know all this;
but, being an intelligent man,
he proceeds to study the behavior
of things around him.

We watch him from the outside,
but he cannot see us.

We notice that
he has a tray in his hands.
And of course we know that
the tray shares the motion of
everything in the box,

101

and therefore remains
relatively at rest to him —
namely, in his hands.
But he does not think of it in
this way;
to him, everything is actually
at rest.

Suddenly he lets go the tray.
Now we know that the tray will
continue to move upward with
CONSTANT velocity;*
and, since we also know that the box
is moving upwards with
an ACCELERATION,
we expect that very soon the floor
will catch up with the tray
and hit it.
And, of course, we see this
actually happen.
Mr. *K* also sees it happen,
but explains it differently, —
he says that everything was still
until he let go the tray,
and then the tray FELL and
hit the floor;
and *K* attributes this to
"A force of gravity."
Now *K* begins to study this "force."
He finds that there is an attraction
between every two bodies,

*Any moving object CONTINUES to move
with CONSTANT speed in a
STRAIGHT LINE, due to inertia,
unless it is stopped by
some external force,
like friction, for example.

and its strength is proportional to
their "gravitational masses,"
and varies inversely as the
square of the distance between them.

He also makes other experiments,
studying the behavior of bodies
pulled along a smooth table top,
and finds that different bodies offer
different degrees of resistance to
this pull,
and he concludes that the resistance
is proportional to the
"inertial mass" of a body.

And then he finds that
ANY object which he releases
FALLS with the SAME acceleration,
and therefore decides that
the gravitational mass and
the inertial mass of a body
are proportional to each other.

In other words, he explains the fact
that all bodies fall with the
SAME acceleration,
by saying that the force of gravity
is such that
the greater the resistance to motion
which a body has,
the harder gravity pulls it,
and indeed this increased pull
is supposed to be
JUST BIG ENOUGH TO OVERCOME
the larger resistance,
and thus produce
THE SAME ACCELERATION IN ALL BODIES!
Now, if Mr. K is a very intelligent

Newtonian physicist,
he says,
"How strange that these two distinct
properties of a body should
always be exactly proportional
to each other.
But experimental facts show
this accident to be true,
and experiments cannot be denied."
But it continues to worry him.

On the other hand,
if K is an Einsteinian relativist,
he reasons entirely differently:
"There is nothing absolute about
my way of looking at phenomena.
Mr. K', outside,
(he means us),
may see this entire room moving
upward with an acceleration,
and attribute all these happenings
to this motion
rather than to
a force of gravity
as I am doing.
His explanation and mine
are equally good,
from our different viewpoints."

This is what Einstein called
the Principle of Equivalence.

Relativist K continues:
"Let me try to see things from
the viewpoint of
my good neighbor, K',
though I have never met him.
He would of course see

the floor of this room come up and
hit ANY object which I might release,
and it would therefore seem
ENTIRELY NATURAL to him
for all objects released
from a given height
at a given time
to reach the floor together,
which of course is actually the case.
Thus, instead of finding out by
long and careful EXPERIMENTATION
that
the gravitational and inertial masses
are proportional,
as my Newtonian ancestors did,
he would predict A PRIORI
that this MUST be the case.
And so,
although the facts are explainable
in either way,
K'''s point of view is
the simpler one,
and throws light on happenings which
I could acquire only by
arduous experimentation, —
if I were not a relativist and hence
quite accustomed to give
equal weight to
my neighbor's viewpoint!"

Of course as we have told the story,
we know that K' is really right:
But remember that
in the actual world
we do not have this advantage:
We cannot "know" which of the two
explanations is "really" correct.

But, since they are EQUIVALENT,
we may select the simpler one,
as Einstein did.

Thus we already see
an advantage in
Einstein's Principle of Equivalence.
And,
as we said in Chapter X,
this is only the beginning,
for it led to his
new Law of Gravitation which
RETAINED ALL THE MERITS OF
NEWTON'S LAW,
and
has additional NEW merits which
Newton's Law did not have.

As we shall see.

XII. A NON-EUCLIDEAN WORLD.

Granting, then,
the Principle of Equivalence,
according to which Mr. K may replace
the idea of a "force of gravity" by
a "fictitious force" due to motion,*
the next question is:
"How does this help us to derive
a new Law of Gravitation?"
In answer to which
we ask the reader to recall
a few simple things which
he learned in elementary physics in
high school:

*The idea of a "fictitious force"
   due to motion
   is familiar to everyone
   in the following example:
   Any youngster knows that
   if he swings a pail full of water
   in a vertical plane
   WITH SUFFICIENT SPEED,
   the water will not fall out of the pail,
   even when the pail is
   actually upside down!
   And he knows that
   the centrifugal "force"
   is due to the motion only;
   since,
   if he slows down the motion,
   the water WILL fall out
   and give him a good dousing.

If a force acts on a moving object
at an angle to this motion,
it will change the course of the object,
and we say that
the body has acquired an
ACCELERATION,
even though its speed may have
remained unchanged!
This can best be seen with the aid of
the following diagram:

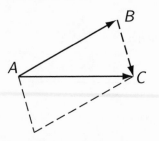

If *AB* represents the original velocity
(both in magnitude and direction)
and if the next second
the object is moving with a velocity
represented by *AC*,
due to the fact that
some force (like the wind)
pulled it out of its course,
then obviously

BC must be the velocity which
had to be "added" to *AB*
to give the "resultant" *AC*,
as any aviator, or even
any high school student, knows from
the "Parallelogram of forces."
Thus *BC* is the difference between
the two velocities, *AC* and *AB*.
And, since
ACCELERATION is defined as
the change in velocity, each second,
then *BC* is the acceleration,
even if *AB* and *AC* happen to be
equal in length, —
that is,
even if the speed of the object
has remained unchanged;*
the very fact that it has merely
changed in DIRECTION
shows that there is an ACCELERATION!
Thus,
if an object moves in a circle,
with uniform speed,
it is moving with
an acceleration since
it is always changing its direction.

Now imagine a physicist who
lives on a disc which
is revolving with constant speed!
Being a physicist,
she is naturally curious about the world,
and wishes to study it,
even as you and I.
And, even though we tell her that

---

*This distinction between "speed" and "velocity"
is discussed on page 128.

she is moving with an acceleration,
she, being a democrat and a relativist,
insists that she can formulate
the laws of the universe
"WITH EQUAL RIGHT AND
EQUAL SUCCESS";
and therefore claims that
she is not moving at all
but is merely in an environment in which
a "force of gravity" is acting
(Have you ever been on a revolving disc
and felt this "force"?!).

Let us now watch her
tackle a problem:
We see her become interested in circles:
She wants to know whether
the circumference of two circles
are in the same ratio as their radii.
She draws two circles,
a large one and a small one
(concentric with
the axis of revolution of the disc)
and proceeds to measure
their radii and circumferences.
When she measures the larger circumference,
we know,
from a study of
The Special Theory of Relativity*
that she will get a different value
from the one WE should get
(not being on the revolving disc);
but this is not the case with
her measurement of the radii,
since the shrinkage in length,
described in the Special Theory,

*See Part I of this book.

takes place only
IN THE DIRECTION OF MOTION,
and not in a direction which is
PERPENDICULAR to the direction of motion
(as a radius is).
Furthermore, when she measures
the circumference of the small circle,
her value is not very different from ours
since the speed of rotation is small
around a small circle,
and the shrinkage is therefore
negligible.
And so, finally, it turns out that
she finds that the circumferences
are NOT in the same ratio as the radii!
Do we tell her that she is wrong?
that this is not according to Euclid?
and that she is a fool for trying
to study Physics on a revolving disc?
Not at all!
On the contrary,
being modern relativists, we say
"That is quite all right, neighbor,
you are probably no worse than we are,
you don't have to use Euclidean geometry if
it does not work on a revolving disc,
for now there are
non-Euclidean geometries which are
exactly what you need —
Just as we would not expect
Plane Trigonometry to work on
a large portion of the earth's surface
for which we need
Spherical Trigonometry,
in which
the angle-sum of a triangle
is NOT 180°,

as we might naively demand after
a high school course in
Euclidean plane geometry.

In short,
instead of considering the disc-world
as an accelerated system,
we can,
by the Principle of Equivalence,
regard it as a system in which
a "force of gravity" is acting,
and, from the above considerations,
we see that
in a space having such a
gravitational field,
Non-Euclidean geometry,
rather than Euclidean,
is applicable.

We shall now illustrate
how the geometry of
a surface or a space may be studied.
This will lead to
the mathematical consideration of
Einstein's Law of Gravitation
and its consequences.

XIII. THE STUDY OF SPACES.

Let us consider first
the familiar Euclidean plane.
Everyone knows that
for a right triangle on such a plane
the Pythagorean theorem holds:
Namely,

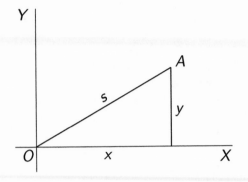

that

$$s^2 = x^2 + y^2$$

Conversely,
it is true that
IF the distance between two points
on a surface
is given by

$$s^2 = x^2 + y^2 \qquad (1)$$

THEN
the surface MUST BE
A EUCLIDEAN PLANE.

Furthermore,
it is obvious that
the distance from O to A
ALONG THE CURVE:

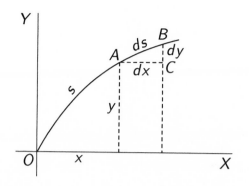

is no longer
the hypotenuse of a right triangle,
and of course
we CANNOT write here $s^2 = x^2 + y^2$!

If, however,
we take two points, A and B,
sufficiently near together,
the curve AB is so nearly
a straight line,
that we may actually regard
ABC as a little right triangle
in which the Pythagorean theorem
does hold.

Only that here
we shall represent its three sides
by ds, dx, and dy,
as is done in
the differential calculus,
to show that
the sides are small.

So that here we have

$$ds^2 = dx^2 + dy^2 \qquad (2)$$

Which still has the form of (1)
and is characteristic of
the Euclidean plane.
It will be found convenient
to replace x and y
by x_1 and x_2, respectively,
so that (2) may be written

$$ds^2 = dx_1^2 + dx_2^2. \qquad (3)$$

Now what is the corresponding situation
on a non-Euclidean surface,
such as
the surface of a sphere, for example?

Let us take
two points on this surface, A and B,
designating the position of each
by its latitude and longitude:

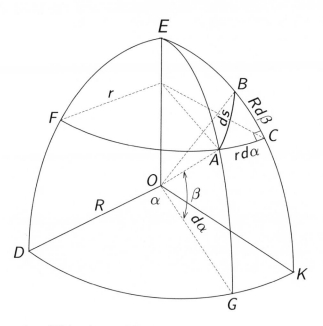

Let *DE* be the meridian
from which
longitude is measured —
the Greenwich meridian.
And let *DK* be a part of the equator,
and *E* the north pole.
Then the longitude and latitude of *A*
are, respectively,
the number of degrees in
the arcs *AF* and *AG*,
(or in the
corresponding central angles, α and β).
Similarly,

the longitude and latitude of B
are, respectively,
the number of degrees in
the arcs CF and BK.

The problem again is
to find the distance
between A and B.
If the triangle ABC is
sufficiently small,
we may consider it to lie
on a Euclidean plane which
practically coincides with
the surface of the sphere in
this little region,
and the sides of the triangle ABC
to be straight lines
(as on page 115).
Then,
since the angle at C
is a right angle,
we have

$$\overline{AB}^2 = \overline{AC}^2 + \overline{BC}^2. \qquad (4)$$

And now let us see
what this expression becomes
if we change
the Cartesian coordinates in (4)
(in the tangent Euclidean plane)
to the coordinates known as
longitude and latitude
on the surface of the sphere.

Obviously AB
has a perfectly definite length
irrespective of

which coordinate system we use;
but AC and BC,
the Cartesian coordinates in
the tangent Euclidean plane
may be transformed into
longitude and latitude on
the surface of the sphere, thus:
let r be
the radius of the latitude circle FAC,
and R the radius of the sphere.
Then*

$$AC = r \cdot d\alpha.$$

Similarly

$$BC = R \cdot d\beta.$$

Therefore, substituting in (4),
we have

$$ds^2 = r^2 d\alpha^2 + R^2 d\beta^2. \tag{5}$$

And, replacing α by x_1, and β by x_2,
this may be written

$$ds^2 = r^2 dx_1^2 + R^2 dx_2^2. \tag{6}$$

A comparison of (6) and (3)
will show that

*any high school student knows that if x represents the length of an arc, and if θ is the number of radians in it, then

$$x/\theta = 2\pi r/2\pi$$

And therefore

$$x = r\theta.$$

PAUL DRY BOOKS, INC.

1616 WALNUT STREET, SUITE 808

PHILADELPHIA, PA 19103

At Paul Dry Books our aim is to publish lively books "to awaken, delight, & educate," and to spark conversation among friends. Our titles include works of fiction and nonfiction—biography, memoirs, history, and essays. We also publish the Nautilus Series for Young Adults, great writing for avid readers of all ages.

To receive our catalog, return this card or e-mail us at pdb@pauldrybooks.com. You can see excerpts, reviews, news about our authors, and articles about us and our books at www.pauldrybooks.com. If you like our books, tell your friends about them.

Name

Address

City, State, Zip Code

Email (the best way for us to communicate with each other)

Book title

BOOKS TO
AWAKEN,
DELIGHT,
&EDUCATE

on the sphere,
the expression for ds^2
is not quite so simple
as it was on the Euclidean plane.

The question naturally arises,
does this distinction between
a Euclidean and a non-Euclidean surface
always hold,
and is this a way
to distinguish between them?

That is,
if we know
the algebraic expression which represents
the distance between two points
which actually holds
on a given surface,
can we then immediately decide
whether the surface
is Euclidean or not?
Or does it perhaps depend upon
the coordinate system used?

To answer this,
let us go back to the Euclidean plane,
and use oblique coordinates
instead of the more familiar
rectangular ones
thus:

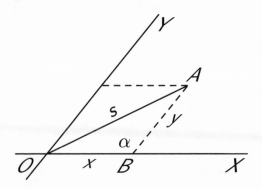

The coordinates of the point A
are now represented by
x and y
which are measured
parallel to the X and Y axes,
and are now
NOT at right angles to each other.

Can we now find
the distance between O and A
using these oblique coordinates?
Of course we can,
for,
by the well-known
Law of Cosines in Trigonometry,
we can represent
the length of a side of a triangle

lying opposite an obtuse angle,
by:

$$s^2 = x^2 + y^2 - 2xy \cos \alpha.$$

Or, for a very small triangle,

$$ds^2 = dx^2 + dy^2 - 2dx\,dy \cos \alpha.$$

And, if we again
replace x and y
by x_1 and x_2, respectively,
this becomes

$$ds^2 = dx_1^2 + dx_2^2 - 2dx_1 \cdot dx_2 \cdot \cos \alpha. \qquad (7)$$

Here we see that
even on a Euclidean plane,
the expression for ds^2
is not as simple as it was before.

And, if we had used
polar coordinates
on a Euclidean plane,
we would have obtained*

$$ds^2 = dr^2 + r^2 d\theta^2$$

or

$$ds^2 = dx_1^2 + x_1^2 dx_2^2. \qquad (8)$$

*See page 124.

123

The reader should verify this,
remembering that
the polar coordinates of point P

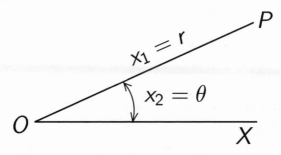

are

(1) its distance, x_1, from a fixed point, O,

(2) the angle, x_2, which OP makes with a fixed line OX.

Then (8) is obvious from
the following figure:

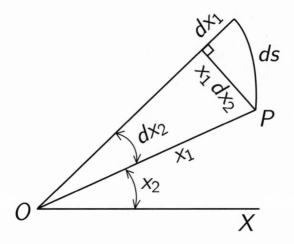

Hence we see that
the form of the expression for ds^2
depends upon BOTH
(a)   the KIND OF SURFACE
       we are dealing with,
and
(b)   the particular
       COORDINATE SYSTEM.

We shall soon see that
whereas
a mere superficial inspection
of the expression for ds^2
is not sufficient
to determine the kind of surface
we are dealing with,
a DEEPER examination
of the expression
DOES help us to know this.
For this deeper examination
we must know
how,
from the expression for ds^2,
to find
the so-called "CURVATURE TENSOR"
of the surface.

And this brings us to
the study of tensors:

What are tensors?
Of what use are they?
and HOW are they used?

Let us see.

XIV. WHAT IS A TENSOR?

The reader is no doubt familar
with the words "scalar" and "vector."
A scalar is a quantity which
has magnitude only,
whereas
a vector has
both magnitude and direction.

Thus,
if we say that
the temperature at a certain place
is 70° Fahrenheit,
there is obviously NO DIRECTION
to this temperature,
and hence
TEMPERATURE is a SCALAR.
But
if we say that
an airplane has gone
one hundred miles east,
obviously its displacement
from its original position
is a VECTOR,
whose MAGNITUDE is 100 miles,
and whose DIRECTION is EAST.

Similarly,
a person's AGE is a SCALAR,
whereas

the VELOCITY with which an object moves
is a VECTOR,*
and so on;
the reader can easily
find further examples
of both scalars and vectors.

We shall now discuss
some quantities
which come up in our experience
and which are
neither scalars nor vectors,
but which are called
TENSORS.
And,
when we have illustrated and defined these,
we shall find that
a SCALAR is a TENSOR whose RANK is ZERO,
and
a VECTOR is a TENSOR whose RANK is ONE,
and we shall see what is meant by
a TENSOR of RANK TWO, or THREE, etc.
Thus "TENSOR" is a more inclusive term,

*A distinction is often made between
"speed" and "velocity" —
the former is a SCALAR, the latter a VECTOR.
Thus when we are interested ONLY in
HOW FAST a thing is moving,
and do not care about its
DIRECTION of motion,
we must then speak of its SPEED,
but if we are interested ALSO in its
DIRECTION,
we must speak of its VELOCITY.
Thus the SPEED of an automobile
would be designated by
"Thirty miles an hour,"
but its VELOCITY would be
"Thirty miles an hour EAST."

of which "SCALAR" and "VECTOR" are
SPECIAL CASES.

Before we discuss
the physical meaning of
a tensor of rank two,
let us consider
the following facts about vectors.

Suppose that we have
any vector, *AB*, in a plane,
and suppose that
we draw a pair of rectangular axes,
X and *Y*,
thus:

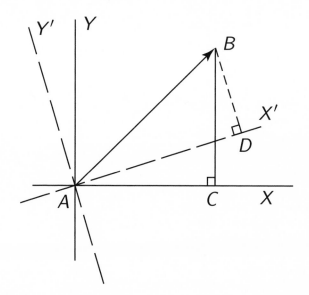

Drop a perpendicular *BC*
from *B* to the *X*-axis.
Then we may say that
AC is the *X*-component of *AB*,
and *CB* is the *Y*-component of *AB*;
for,
as we know from
the elementary law of
"The parallelogram of forces,"
if a force *AC* acts on a particle
and *CB* also acts on it,
the resultant effect is the same
as that of a force *AB* alone.
And that is why
AC and *CB* are called
the "components" of *AB*.
Of course if we had used
the dotted lines as axes instead,
the components of *AB*
would now be *AD* and *DB*.
In other words,
the vector *AB* may be broken up
into components
in various ways,
depending on our choice of axes.

Similarly,
if we use THREE axes in SPACE
rather than two in a plane,
we can break up a vector
into THREE components
as shown
in the diagram
on page 131.

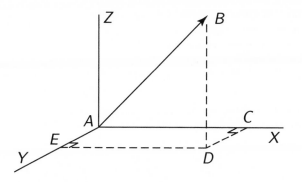

By dropping the perpendicular *BD*
from *B* to the *XY*-plane,
and then drawing
the perpendiculars *DC* and *DE*
to the *X* and *Y* axes, respectively,
we have the three components of *AB*,
namely,
AC, *AE*, and *DB*;
and, as before,
the components depend upon
the particular choice of axes.

Let us now illustrate
the physical meaning
of a tensor of rank two.

Suppose we have a rod
at every point of which
there is a certain strain
due to some force acting on it.
As a rule the strain

is not the same at all points,
and, even at any given point,
the strain is not the same in
all directions.*
Now, if the STRESS at the point *A*
(that is, the FORCE causing the strain at *A*)
is represented
both in magnitude and direction
by *AB*

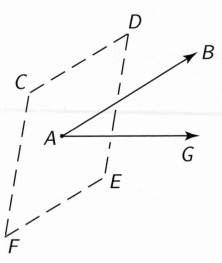

*When an object finally breaks
  under a sufficiently great strain,
  it does not fly into bits
  as it would do if
  the strain were the same
  at all points and in all directions,
  but breaks along certain lines
  where, for one reason of another,
  the strain is greatest.

and if we are interested to know
the effect of this force upon
the surface *CDEF* (through *A*),
we are obviously dealing
with a situation which depends
not on a SINGLE vector,
but on TWO vectors:
Namely,
one vector, *AB*,
which represents the force in question,
and another vector
(call it *AG*),
whose direction will indicate
the ORIENTATION of this surface *CDEF*,
and whose magnitude will represent
the AREA of *CDEF*.

In other words,
the effect of a force upon a surface
depends NOT ONLY on the force itself
but ALSO on the
size and orientation of the surface.

Now, how can we indicate
the orientation of a surface
by a line?
If we draw a line through *A*
in the plane *CDEF*,
obviously we can draw this line
in many different directions,
and there is no way
of choosing one of these
to represent the orientation of this surface.
BUT,
if we take a line through *A*
PERPENDICULAR to the plane *CDEF*,
such a line is UNIQUE

and CAN therefore be used
to specify the orientation
of the surface $CDEF$.
Hence, if we draw a vector, AG,
in a direction perpendicular to $CDEF$
and of such a length that
it represents the magnitude of
the area of $CDEF$,
then obviously
this vector AG
indicates clearly
both the SIZE and the ORIENTATION
of the surface $CDEF$.

Thus,
the STRESS at A
upon the surface $CDEF$
depends upon the TWO vectors,
AB and AG,
and is called
a TENSOR of RANK TWO.

Let us now find a convenient way
of representing this tensor.
And, in order to do so,
let us consider the stress, F,
upon a small surface, dS,
represented in the following figure
by ABC $(= dS)$.
Now if OG, perpendicular to ABC,
is the vector which represents
the size and orientation of ABC,
then,

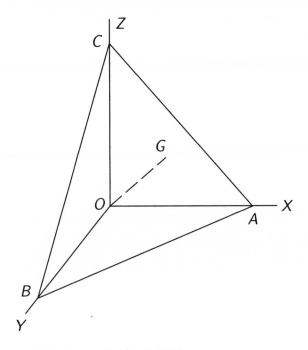

it is quite easy to see (page 136)

that the X-component of OG
represents in magnitude and direction
the projection OBC of ABC upon the YZ-plane.
And similarly,
the Y and Z components of OG
represent the projections
OAC and OAB, respectively.

To show that *OK* represents *OBC*
both in magnitude and direction:

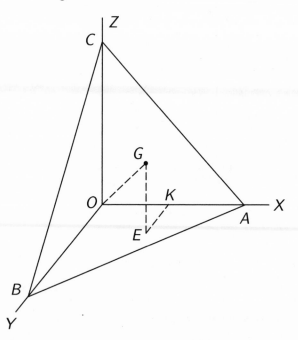

That it does so in direction
is obvious.
since *OK* is perpendicular to *OBC* (see p. 134).
As regards the magnitude
we must now show that

$$\frac{OK}{OG} = \frac{OBC}{ABC}:$$

(a) Now *OBC* = *ABC* × cos of the dihedral angle
between *ABC* and *OBC*
(since the area of the projection
of a given surface
is equal to
the area of the given surface multiplied by

136

the cosine of the dihedral angle
between the two planes).
But this dihedral angle equals angle GOK
since OG and OK are respectively
perpendicular to ABC and OBC,
and $\cos \angle GOK$ is OK/OG.
Substitution of this in (a)
gives the required

$$\frac{OBC}{ABC} = \frac{OK}{OG}.$$

Now, if the force F,
which is itself a vector,
acts on ABC,
we can examine its total effect
by considering separately
the effects of its three components

$$f_x, \ f_y, \ \text{and} \ f_z$$

upon EACH of the three projections

$$OBC, \ OAC, \ \text{and} \ OAB.$$

Let us designate these projections
by dS_x, dS_y, and dS_z, respectively.

Now,
since f_x
(which is the X-component of F)
acts upon EACH one of the three
above-mentioned projections,
let us designate the pressure
due to this component alone
upon the three projections
by

$$p_{xx}, \ p_{xy}, \ p_{xz},$$

respectively.
We must emphasize
the significance of this notation:
In the first place,

the reader must distinguish between
the "pressure" on a surface
and the "force" acting on the surface.
The "pressure" is
the FORCE PER UNIT AREA.
So that
the TOTAL FORCE is obtained by
MULTIPLYING
the PRESSURE by the AREA of the surface.
Thus the product

$$p_{xx} \cdot dS_x$$

gives the force acting upon
the projection dS_x
due to the action of f_x ALONE.
Note the DOUBLE subscripts in

$$p_{xx}, \ p_{xy}, \ p_{xz} \ :$$

The first one obviously refers to the fact
that
these three pressures all emanate
from the component f_x alone;
whereas,
the second subscript designates
the particular projection upon which
the pressure acts.
Thus p_{xy} means
the pressure due to f_x
upon the projection dS_y,
Etc.
It follows therefore that

$$f_x = p_{xx} \cdot dS_x + p_{xy} \cdot dS_y + p_{xz} \cdot dS_z.$$

And, similarly,

$$f_y = p_{yx} \cdot dS_x + p_{yy} \cdot dS_y + p_{yz} \cdot dS_z$$

and

$$f_z = p_{zx} \cdot dS_x + p_{zy} \cdot dS_y + p_{zz} \cdot dS_z.$$

138

Hence the TOTAL STRESS, F,
on the surface dS,
is
$$F = f_x + f_y + f_z$$
or
$$\begin{aligned} F = {} & p_{xx} \cdot dS_x + p_{xy} \cdot dS_y + p_{xz} \cdot dS_z \\ & + p_{yx} \cdot dS_x + p_{yy} \cdot dS_y + p_{yz} \cdot dS_z \\ & + p_{zx} \cdot dS_x + p_{zy} \cdot dS_y + p_{zz} \cdot dS_z. \end{aligned}$$

Thus we see that
stress is not just a vector,
with three components in
three-dimensional space (see p. 130)
but has NINE components
in THREE-dimensonal space.
Such a quantity is called
A TENSOR OF RANK TWO.

For the present
let this illustration of a tensor suffice:
Later we shall give a precise definition.

It is obvious that
if we were dealing with a plane
instead of with
three-dimensional space,
a tensor of rank two would then have
only FOUR components instead of nine,
since each of the two vectors involved
has only two components in a plane,
and therefore,
there would now be only
2×2 components for the tensor
instead of 3×3 as above.

And, in general,
if we are dealing with
n-dimensional space,

a tensor of rank two
has n^2 components
which are therefore conveniently written
in a SQUARE array
as was done on page 139.
Whereas,
in n-dimensional space,
a VECTOR has only n components:
Thus,
a VECTOR in a PLANE
has TWO components;
in THREE-dimensional space it has
THREE components;
and so on.

Hence,
the components of a VECTOR
are therefore written
in a SINGLE ROW;
instead of in a SQUARE ARRAY
as in the case of a TENSOR of RANK TWO.

Similarly,
in n-dimensional space
a TENSOR of rank THREE has n^3 components,
and so on.

To sum up:

In n-dimensional space,
a VECTOR has n components,
a TENSOR of rank TWO has n^2 components,
a TENSOR of rank THREE has n^3 components,
and so on.

The importance of tensors
in Relativity
will become clear
as we go on.

XV. THE EFFECT ON TENSORS OF A CHANGE IN THE COORDINATE SYSTEM.

In Part I of this book (page 61)
we had occasion to mention
the fact that
the coordinates of the point A

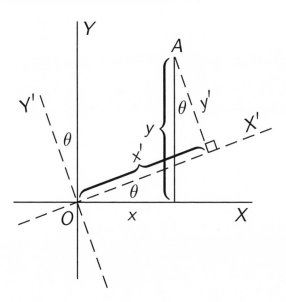

in the unprimed coordinate system
can be expressed in terms of
its coordinates in the
primed coordinate system
by the relationships

$$\left. \begin{array}{l} x = x' \cos\theta - y' \sin\theta \\ y = x' \sin\theta + y' \cos\theta \end{array} \right\} \qquad (9)$$

as is known to any young student of
elementary analytical geometry.

Let us now see
what effect this change in
the coordinate system
has
upon a vector and its components.

Call the vector ds,
and let dx and dy represent
its components in the UNPRIMED SYSTEM,
and dx' and dy'
its components in the PRIMED SYSTEM
as shown on page 143.

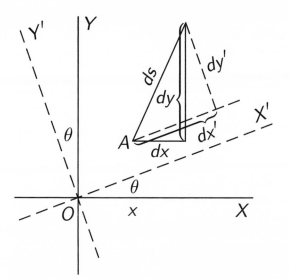

Obviously *ds* by itself
is not affected by the change
of coordinate system,
but the COMPONENTS of *ds*
in the two systems
are DIFFERENT,
as we have already pointed out
on page 130.

Now if the coordinates of point A
are *x* and *y* in one system

and x' and y' in the other,
the relationhip between
these four quantities
is given by equations (9) on p. 142.
And now, from these equations,
we can, by differentiation,*
find the relationships between
dx and dy
and
dx' and dy'.

It will be noticed,
in equations (9),
that
x depends on BOTH x' and y',
so that any changes in x' and y'
will BOTH affect x.
Hence the TOTAL change in x,
namely dx,
will depend upon TWO causes:

(a) Partially upon the change in x',
      namely dx',
and
(b) Partially upon the change in y',
      namely dy'.

Before writing out these changes,
it will be found more convenient
to solve (9) for x' and y'
in terms of x and y.†

  *See any calculus textbook. [9]

  †Assuming of course that the
    determinant of the coefficients in (9)
    is not zero. [10]

In other words,
to express the
NEW, primed coordinates, x' and y',
in terms of the
OLD, original ones, x and y,
rather than the other way around.

This will of course give us

$$\left.\begin{array}{l} x' = ax + by \\ y' = cx + dy \end{array}\right\} \qquad (10)$$

where a, b, c, d are functions of θ.

It will be even better
to write (10) in the form:

$$\left.\begin{array}{l} x'_1 = a_{11}x_1 + a_{12}x_2 \\ x'_2 = a_{21}x_1 + a_{22}x_2 \end{array}\right\} \qquad (11)$$

using x_1 and x_2 instead of x and y,
(and of course x'_1 and x'_2 instead of
x' and y');
and putting different subscripts
on the single letter a,
instead of using
four different letters: a, b, c, d.
The advantage of this notation is
not only that we can
easily GENERALIZE to n dimensions
from the above
two-dimensional statements,
but,
as we shall see later,
this notation lends itself to
a beautifully CONDENSED way of
writing equations,
which renders them
very EASY to work with.

Let us now proceed with
the differentiation of (11):
we get

$$\left.\begin{array}{l} dx_1' = a_{11}dx_1 + a_{12}dx_2 \\ dx_2' = a_{21}dx_1 + a_{22}dx_2 \end{array}\right\} \qquad (12)$$

The MEANING of the a's in (12)
should be clearly understood:
Thus a_{11} is
the change in x_1' due to
A UNIT CHANGE in x_1,
so that
when it is multiplied by
the total change in x_1, namely dx_1,
we get
THE CHANGE IN x_1' DUE TO
THE CHANGE IN x_1 ALONE.
And similarly in $a_{12}dx_2$,
a_{12} represents
the change in x_1' PER UNIT CHANGE in x_2,
and therefore
the product of a_{12} and
the total change in x_2, namely dx_2,
gives
THE CHANGE IN x_1' DUE TO
THE CHANGE IN x_2 ALONE.

Thus
the TOTAL CHANGE in x_1'
is given by

$$a_{11}dx_1 + a_{12}dx_2,$$

just as
the total cost of
a number of apples and oranges
would be found
by multiplying the cost of
ONE APPLE
by the total number of apples,

and ADDING this result
to a similar one
for the oranges.

And similarly for dx_2' in (12).

We may therefore
replace a_{11} by $\partial x_1'/\partial x_1$
a symbol which represents
the partial change in x_1'
per unit change in x_1,
and is called*
the "partial derivative of x_1'
with respect to x_1."

Similarly,

$$a_{12} = \frac{\partial x_1'}{\partial x_2}, \quad a_{21} = \frac{\partial x_2'}{\partial x_1}, \quad a_{22} = \frac{\partial x_2'}{\partial x_2}.$$

And we may therefore rewrite (12)
in the form

$$\left. \begin{array}{l} dx_1' = \dfrac{\partial x_1'}{\partial x_1} \cdot dx_1 + \dfrac{\partial x_1'}{\partial x_2} \cdot dx_2 \\[3mm] dx_2' = \dfrac{\partial x_2'}{\partial x_1} \cdot dx_1 + \dfrac{\partial x_2'}{\partial x_2} \cdot dx_2 \end{array} \right\} \tag{13}$$

But perhaps the reader
is getting a little tired of all this,
and is wondering
what it has to do
with Relativity.

---

*Note that a PARTIAL change
is always denoted by the letter "∂"
in contrast to "d"
which designates a TOTAL change. [11]

To which we may give him
a partial answer now
and hold out hope
of further information
in the remaining chapters.
What we can already say is that
since General Relativity is concerned with
finding the laws of the physical world
which hold good for ALL observers,*
and since various observers
differ from each other,
as physicists,
only in that they
use different coordinate systems,
we see then
that Relativity is concerned
with finding out those things
which remain INVARIANT
under transformations of
coordinate systems.

Now, as we saw on page 143,
a vector is such an INVARIANT,
and, similarly,
tensors in general
are such INVARIANTS,
so that the business of the physicist
really becomes
to find out
which physical quantities
are tensors,
and are therefore
the "facts of the universe,"
since they hold good
for all observers.

*See p. 96.

Besides,
as we promised on page 125,
we must explain the meaning of
"curvature tensor,"
since it is this tensor
which CHARACTERIZES a space.

And then
with the aid of the curvature tensor of
our four-dimensional world of events,*
we shall find out
how things move in this world —
what paths the planets take,
and in what path
a ray of light travels
as it passes near the sun,
and so on.

And of course
these are all things which
can be
VERIFIED BY EXPERIMENT.

XVI. A VERY HELPFUL SIMPLIFICATION.

Before we go any further
let us write equations (13) on page 147
more briefly
thus:

$$dx'_\mu = \sum_\sigma \frac{\partial x'_\mu}{\partial x_\sigma} \cdot dx_\sigma \qquad \left(\begin{matrix} \mu = 1,\ 2. \\ \sigma = 1,\ 2. \end{matrix} \right) \qquad (14)$$

*FOUR-dimensional, since
  each event is characterized by
  its THREE space-coordinates and
  the TIME of its occurrence
  (see Part I of this book, page 58).

150

A careful study of (14) will show

(a) That (14) really contains TWO equations
(although it only looks like one),
since, as we give μ
its possible values, 1 and 2,
we have
dx_1' and dx_2' on the left,
just as we did in (13);

(b) The symbol $\sum\limits_{\sigma}$ means that
when the various values of σ,
namely 1 and 2,
are substituted for σ
(keeping the μ constant in any one equation)
the resulting two terms
must be ADDED together.

Thus, for $\mu = 1$ and $\sigma = 1, 2$,
(14) becomes

$$dx_1' = \frac{\partial x_1'}{\partial x_1} \cdot dx_1 + \frac{\partial x_1'}{\partial x_2} \cdot dx_2,$$

just like the FIRST equation in (13),
and, similarly,
by taking $\mu = 2$,
and again "summing on the σ's,"
since that is what $\sum\limits_{\sigma}$ tells us to do,
we get

$$dx_2' = \frac{\partial x_2'}{\partial x_1} \cdot dx_1 + \frac{\partial x_2'}{\partial x_2} \cdot dx_2,$$

which is the SECOND equation in (13).
Thus we see that
(14) includes all of (13).

A still further abbreviation
is introduced by omitting
the symbol $\sum\limits_{\sigma}$

WITH THE UNDERSTANDING THAT
WHENEVER A SUBSCRIPT OCCURS TWICE
IN A SINGLE TERM
(as, for example, σ
in the right-hand member of (14)),
it will be understood that
a SUMMATION is to be made
ON THAT SUBSCRIPT.
Hence we may write (14) as follows:

$$dx'_\mu = \frac{\partial x'_\mu}{\partial x_\sigma} \cdot dx_\sigma \qquad \begin{pmatrix} \mu = 1, 2 \\ \sigma = 1, 2 \end{pmatrix} \qquad (15)$$

in which we shall know
that the presence of the TWO σ's
in the term on the right,
means that $\underset{\sigma}{\Sigma}$ is understood.

And now, finally,
since dx_1 and dx_2
are the components of ds in the
UNPRIMED system
let us represent them more briefly by

$$A^1 \text{ and } A^2$$

respectively.
The reader must NOT confuse
these SUPERSCRIPTS
with EXPONENTS —
thus A^2 is not the "square of" A,
but the superscript serves merely
the same purpose as a
SUBSCRIPT,
namely,
to distinguish the components
from each other.
Just why we use
SUPERSCRIPTS instead of subscripts
will appear later (p. 172).

152

And the components of *ds*
in the PRIMED coordinate system
will now be written
$$A'^1 \text{ and } A'^2.$$

Thus (15) becomes

$$A'^\mu = \frac{\partial x'_\mu}{\partial x_\sigma} \cdot A^\sigma. \tag{16}$$

And so,
if we have a certain vector A^σ,
that is,
a vector whose components are
A^1 and A^2
in a certain coordinate system,
and if we change to
a new coordinate system
in accordance with
the transformation represented by (11) on page 145,
then
(16) tells us what will be
the components of this same vector
in the new (PRIMED) coordinate system.

Indeed, (15) or (16) represents
the change in the components
of a vector
NOT ONLY for the change given in (11),
but for ANY transformation
of coordinates:*
Thus
suppose x_σ are the coordinates of
a point in one coordinate system,

*Except only that
the values of (x_σ) and (x'_μ) must be in
one-to-one correspondence.

and suppose that

$$x_1' = f_1(x_1, x_2, \ldots) = f_1(x_\sigma)$$
$$x_2' = f_2(x_\sigma)$$

etc.

Or, representing this entire
set of equations by

$$x_\mu' = f_\mu(x_\sigma)$$

where the f's represent
any functions whatever,
then, obviously

$$dx_1' = \frac{\partial f_1}{\partial x_1} \cdot dx_1 + \frac{\partial f_1}{\partial x_2} \cdot dx_2 + \ldots$$

or, since $f_1 = x_1'$,

$$dx_1' = \frac{\partial x_1'}{\partial x_1} \cdot dx_1 + \frac{\partial x_1'}{\partial x_2} \cdot dx_2 + \ldots$$

etc.

Hence

$$dx_\mu' = \frac{\partial x_\mu'}{\partial x_\sigma} \cdot dx_\sigma \quad \text{or} \quad A'^\mu = \frac{\partial x_\mu'}{\partial x_\sigma} \cdot A^\sigma$$

gives the manner of transformation
of the vector dx_σ to
ANY other coordinate system
(see the only limitation
mentioned in the footnote on
page 154).

And in fact
ANY set of quantities which
transforms according to (16) is
DEFINED TO BE A VECTOR,
or rather,
A CONTRAVARIANT VECTOR —
the meaning of "CONTRAVARIANT"

155

will appear later (p. 172).

The reader must not forget that
whereas the separate components in
the two coordinate systems are
different,
the vector itself is
an INVARIANT under the
transformation of coordinates
(see page 143).
It should be noted further that
(16) serves not only to represent
a two-dimensional vector,
but may represent
a three- or four- or
n-dimensional vector,
since all that is necessary is
to indicate the number of values that
μ or σ may take.
Thus, if $\mu = 1, 2$, and $\sigma = 1, 2$,
we have a two-dimensional vector;
but if $\mu = 1, 2, 3$, and $\sigma = 1, 2, 3$,
(16) represents a 3-dimensional vector,
and so on.
For the case $\mu = 1, 2, 3$ and $\sigma = 1, 2, 3$,
(16) obviously represents
THREE EQUATIONS in which
the right-hand members
each have THREE terms:

$$A'^1 = \frac{\partial x_1'}{\partial x_1} \cdot A^1 + \frac{\partial x_1'}{\partial x_2} \cdot A^2 + \frac{\partial x_1'}{\partial x_3} \cdot A^3$$

$$A'^2 = \frac{\partial x_2'}{\partial x_1} \cdot A^1 + \frac{\partial x_2'}{\partial x_2} \cdot A^2 + \frac{\partial x_2'}{\partial x_3} \cdot A^3$$

$$A'^3 = \frac{\partial x_3'}{\partial x_1} \cdot A^1 + \frac{\partial x_3'}{\partial x_2} \cdot A^2 + \frac{\partial x_3'}{\partial x_3} \cdot A^3$$

Similarly we may now give
the mathematical definition of
a tensor of rank two,*
or of any other rank.
Thus
a contravariant tensor of rank two
is defined as follows:

$$A'^{\alpha\beta} = \frac{\partial x'_\alpha}{\partial x_\gamma} \cdot \frac{\partial x'_\beta}{\partial x_\delta} \cdot A^{\gamma\delta} \tag{17}$$

Here, since γ and δ occur TWICE
in the term on the right,
it is understood that
we must SUM for these indices
over whatever range of values they have.
Thus if we are speaking of
THREE DIMENSIONAL SPACE,
we have $\gamma = 1, 2, 3$ and $\delta = 1, 2, 3$.
ALSO $\alpha = 1, 2, 3$ and $\beta = 1, 2, 3$;
But
NO SUMMATION is to be performed
on the α and β
since neither of them occurs
TWICE in a single term;
so that
any particular values of α and β
must be retained throughout ANY ONE equation.

For example,
for the case $\alpha = 1$, $\beta = 2$,

*It will be remembered (see page 128)
 that
 a VECTOR is a TENSOR of RANK ONE.

(17) gives the equation:

$$A'^{12} = \frac{\partial x_1'}{\partial x_1} \cdot \frac{\partial x_2'}{\partial x_1} \cdot A^{11} + \frac{\partial x_1'}{\partial x_1} \cdot \frac{\partial x_2'}{\partial x_2} \cdot A^{12} + \frac{\partial x_1'}{\partial x_1} \cdot \frac{\partial x_2'}{\partial x_3} \cdot A^{13}$$

$$+ \frac{\partial x_1'}{\partial x_2} \cdot \frac{\partial x_2'}{\partial x_1} \cdot A^{21} + \frac{\partial x_1'}{\partial x_2} \cdot \frac{\partial x_2'}{\partial x_2} \cdot A^{22} + \frac{\partial x_1'}{\partial x_2} \cdot \frac{\partial x_2'}{\partial x_3} \cdot A^{23}$$

$$+ \frac{\partial x_1'}{\partial x_3} \cdot \frac{\partial x_2'}{\partial x_1} \cdot A^{31} + \frac{\partial x_1'}{\partial x_3} \cdot \frac{\partial x_2'}{\partial x_2} \cdot A^{32} + \frac{\partial x_1'}{\partial x_3} \cdot \frac{\partial x_2'}{\partial x_3} \cdot A^{33}$$

It will be observed that γ and δ
have each taken on
their THREE possible values: 1, 2, 3,
which resulted in
NINE terms on the right,
whereas
$\alpha = 1$ and $\beta = 2$
have been retained throughout.

And now since α and β
may each have the three values, 1, 2, 3,
there will be NINE such EQUATIONS in all.

Thus (17) represents
nine equations each containing
nine terms on the right,
if we are considering
three-dimensional space.
Obviously for two-dimensional space,
(17) will represent
only four equations each containing
only four terms on the right.
Whereas,
in four dimensions,
as we must have in
Relativity*

*See the footnote on p. 150.

(17) will represent
sixteen equations each containing
sixteen terms on the right.

And, in general,
in n-dimensional space,
a tensor of RANK TWO,
defined by (17),
consists of
n^2 equations, each containing
n^2 terms in the right-hand member.

Similarly,
a contravariant tensor of RANK THREE
is defined by

$$A'^{\alpha\beta\gamma} = \frac{\partial x'_\alpha}{\partial x_\mu} \cdot \frac{\partial x'_\beta}{\partial x_\nu} \cdot \frac{\partial x'_\gamma}{\partial x_\sigma} \cdot A^{\mu\nu\sigma} \qquad (18)$$

and so on.
As before,
the number of equations represented by (18)
and the number of terms on the right in each,
depend upon
the dimensionality of the space in question.

The reader can already appreciate somewhat
the remarkable brevity
of this notation.
But when he will see in the next chapter
how easily such sets of equations
are MANIPULATED,
he will be really delighted,
we are sure of that.

XVII. OPERATIONS WITH TENSORS.

For example,
take the vector (or tensor of rank one) A^α,
having the two components A^1 and A^2
in a plane,
with reference to a given set of axes.
And let B^α be another such vector.
Then, by adding the corresponding components
of A^α and B^α,
we obtain a quantity
also having two components,
namely,

$$A^1 + B^1 \text{ and } A^2 + B^2$$

which may be represented by

$$C^1 \text{ and } C^2,$$

respectively.

Let us now prove
that this quantity
is also a vector:
Since A^α is a vector,
its law of transformation is (see p. 153):

$$A'^\lambda = \frac{\partial x'_\lambda}{\partial x_\alpha} \cdot A^\alpha \qquad (19)$$

Similarly, for B^α:

$$B'^\lambda = \frac{\partial x'_\lambda}{\partial x_\alpha} \cdot B^\alpha. \qquad (20)$$

Taking corresponding components,

we get, in full:

$$A'^1 = \frac{\partial x_1'}{\partial x_1} \cdot A^1 + \frac{\partial x_1'}{\partial x_2} \cdot A^2$$

and

$$B'^1 = \frac{\partial x_1'}{\partial x_1} \cdot B^1 + \frac{\partial x_1'}{\partial x_2} \cdot B^2.$$

The sum of these gives:

$$A'^1 + B'^1 = \frac{\partial x_1'}{\partial x_1} \cdot \left(A^1 + B^1\right) + \frac{\partial x_1'}{\partial x_2} \cdot \left(A^2 + B^2\right).$$

Similarly,

$$A'^2 + B'^2 = \frac{\partial x_2'}{\partial x_1} \cdot \left(A^1 + B^1\right) + \frac{\partial x_2'}{\partial x_2} \cdot \left(A^2 + B^2\right).$$

Both these results are included in:

$$A'^\lambda + B'^\lambda = \frac{\partial x_\lambda'}{\partial x_\alpha} \cdot \left(A^\alpha + B^\alpha\right) \qquad (\lambda, \alpha = 1, 2)$$

Or

$$C'^\lambda = \frac{\partial x_\lambda'}{\partial x_\alpha} \cdot C^\alpha. \qquad (21)$$

Thus we see that
the result is
a VECTOR (see p. 155).

Similarly for tensors of
higher ranks.

Furthermore,
note that (21) may be obtained
QUITE MECHANICALLY
by adding (19) and (20)
AS IF each of these were
A SINGLE equation
containing only
A SINGLE term on the right,

instead of
A SET OF EQUATIONS
EACH CONTAINING
SEVERAL TERMS ON THE RIGHT.

Thus the notation
AUTOMATICALLY takes care that
the corresponding components
shall be properly added.

This is even more impressive
in the case of multiplication.
Thus,
to multiply

$$A'^\lambda = \frac{\partial x'_\lambda}{\partial x_\alpha} \cdot A^\alpha \tag{22}$$

by

$$B'^\mu = \frac{\partial x'_\mu}{\partial x_\beta} \cdot B^\beta \qquad (\lambda, \mu, \alpha, \beta = 1, 2) \tag{23}$$

we write the result immediately:

$$C'^{\lambda\mu} = \frac{\partial x'_\lambda}{\partial x_\alpha} \cdot \frac{\partial x'_\mu}{\partial x_\beta} \cdot C^{\alpha\beta}. \qquad (\lambda, \mu, \alpha, \beta = 1, 2) \tag{24}$$

To convince the reader
that it is quite safe
to write the result so simply,
let us examine (24) carefully
and see whether it really represents
correctly
the result of multiplying (22) by (23).
By "multiplying (22) by (23)"
we mean that
EACH equation of (22) is to be
multiplied by
EACH equation of (23)

in the way in which this would be done
in ordinary algebra.
Thus,
we must first multiply

$$A'^1 = \frac{\partial x'_1}{\partial x_1} \cdot A^1 + \frac{\partial x'_1}{\partial x_2} \cdot A^2$$

by

$$B'^1 = \frac{\partial x'_1}{\partial x_1} \cdot B^1 + \frac{\partial x'_1}{\partial x_2} \cdot B^2.$$

We get,

$$\begin{aligned}
A'^1 B'^1 = &\frac{\partial x'_1}{\partial x_1} \cdot \frac{\partial x'_1}{\partial x_1} \cdot A^1 B^1 + \\
&\frac{\partial x'_1}{\partial x_2} \cdot \frac{\partial x'_1}{\partial x_1} \cdot A^2 B^1 + \\
&\frac{\partial x'_1}{\partial x_1} \cdot \frac{\partial x'_1}{\partial x_2} \cdot A^1 B^2 + \\
&\frac{\partial x'_1}{\partial x_2} \cdot \frac{\partial x'_1}{\partial x_2} \cdot A^2 B^2.
\end{aligned}$$

(25)

Similarly we shall get
three more such equations,
whose left-hand members are,
respectively,

$$A'^1 B'^2, \ A'^2 B'^1, \ A'^2 B'^2,$$

and whose right-hand members
resemble that of (25).

Now, we may obtain (25) from (24)
by taking $\lambda = 1$, $\mu = 1$,
retaining these values throughout,
since no summation is indicated on λ and μ
[that is, neither λ nor μ is repeated
in any one term of (24)].

But since α and β
each OCCUR TWICE
in the term on the right,
they must be allowed to take on
all possible values, namely, 1 and 2,
and SUMMED,
thus obtaining (25),
except that we replace $A^\alpha B^\beta$
by the simpler* symbol $C^{\alpha\beta}$.
Similarly,
by taking $\lambda = 1$, $\mu = 2$ in (24),
and summing on α and β as before,
we obtain another of the equations
mentioned on page 164.

And $\lambda = 2$, $\mu = 1$,
gives the third of these equations;
and finally $\lambda = 2, \mu = 2$
gives the fourth and last.

Thus (24) actually does represent
COMPLETELY
the product of (22) and (23)!

Of course, in three-dimensional space,
(22) and (23) would each represent
THREE equations, instead of two,
each containing
THREE terms on the right, instead of two;
and the product of (22) and (23)

*Note that either $A^\alpha B^\beta$ or $C^{\alpha\beta}$
allows for FOUR components:
Namely, $A^1 B^1$ or C^{11},
$A^1 B^2$ or C^{12},
$A^2 B^1$ or C^{21},
and $A^2 B^2$ or C^{22}.
And hence we may use
$C^{\alpha\beta}$ instead of $A^\alpha B^\beta$.

would then consist of
NINE equations, instead of four,
each containing
NINE terms on the right, instead of four.
But this result
is still represented by (24)!
And, of course, in four dimensions
(24) would represent
SIXTEEN equations, and so on.

Thus the tensor notation enables us
to multiply
WHOLE SETS OF EQUATIONS
containing MANY TERMS IN EACH,
as EASILY as we multiply
simple monomials in elementary algebra!

Furthermore,
we see from (24)
that
the PRODUCT of two tensors
is also a TENSOR (see page 157),
and, specifically, that
the product of two tensors
each of RANK ONE,
gives a tensor of RANK TWO.

In general,
if two tensors of ranks m and n,
respectively,
are multiplied together,
the result is
a TENSOR OF RANK $m + n$.

This process of multiplying tensors
is called
OUTER multiplication,

to distinguish it from
another process known as
INNER multiplication
which is also important
in Tensor Calculus,
and which we shall describe later (page 183).

XVIII. A PHYSICAL ILLUSTRATION.

But first let us discuss
a physical illustration of
ANOTHER KIND OF TENSOR,
A COVARIANT TENSOR:*

Consider an object whose density
is different in different parts of the object.

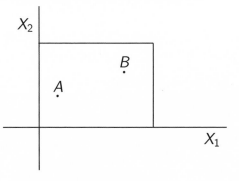

*This is to be distinguished from the
CONTRAVARIANT tensors
discussed on pages 155ff.

We may then speak of
the density at a particular point, A.
Now, density is obviously
NOT a directed quantity,
but a SCALAR (see page 127).
And since the density of the given object
is not uniform throughout,
but varies from point to point,
it will vary as we go from A to B.
So that if we designate by ψ
the density at A,
then

$$\frac{\partial \psi}{\partial x_1} \text{ and } \frac{\partial \psi}{\partial x_2}$$

represent, respectively,
the partial variation of ψ
in the x_1 and x_2 directions.
Thus, although ψ itself is NOT
a DIRECTED quantity,
the CHANGE in ψ DOES depend upon
the DIRECTION
and IS therefore a DIRECTED quantity,
whose components are

$$\frac{\partial \psi}{\partial x_1} \text{ and } \frac{\partial \psi}{\partial x_2}.$$

Now let us see
what happens to this quantity when
the coordinate system is changed (see page 149).

We are seeking to express

$$\frac{\partial \psi}{\partial x_1'}, \frac{\partial \psi}{\partial x_2'} \text{ in terms of } \frac{\partial \psi}{\partial x_1}, \frac{\partial \psi}{\partial x_2}.$$

Now if we have three variables,
say, x, y, and z,

such that y and z depend on x,
it is obvious that
the change in z per unit charge in x,
IF IT CANNOT BE FOUND DIRECTLY,
may be found by
multiplying
the change in y per unit charge in x
by
the change in z per unit change in y,
or,
expressing this in symbols:

$$\frac{dz}{dx} = \frac{dz}{dy} \cdot \frac{dy}{dx}. \tag{26}$$

In our problem above,
we have the following similar situation:
A change in x_1' will affect
BOTH x_1 and x_2 (see p. 145),
and the resulting changes in x_1 and x_2
will affect ψ;
hence

$$\frac{\partial \psi}{\partial x_1'} = \frac{\partial \psi}{\partial x_1} \cdot \frac{\partial x_1}{\partial x_1'} + \frac{\partial \psi}{\partial x_2} \cdot \frac{\partial x_2}{\partial x_1'} \tag{27}$$

Note that here we have TWO terms
on the right
instead of only ONE, as in (26),
since the change in x_1'
affects BOTH x_1 and x_2
and these in turn BOTH affect ψ,
whereas in (26),
a change in x affects y,
which in turn affects z,
and that is all there was to it.
Note also that
the curved "∂" is used throughout in (27)
since all the changes here

are PARTIAL changes
(see footnote on page 147).
And since ψ is influenced also
by a change in x_2',
this influence may be
similarly represented by

$$\frac{\partial \psi}{\partial x_2'} = \frac{\partial \psi}{\partial x_1} \cdot \frac{\partial x_1}{\partial x_2'} + \frac{\partial \psi}{\partial x_2} \cdot \frac{\partial x_2}{\partial x_2'}. \tag{28}$$

And, as before,
we may combine (27) and (28)
by means of the abbreviated notation:

$$\frac{\partial \psi}{\partial x_\mu'} = \frac{\partial \psi}{\partial x_\sigma} \cdot \frac{\partial x_\sigma}{\partial x_\mu'} \qquad (\mu, \sigma = 1, 2) \tag{29}$$

where the occurrence of σ TWICE
in the single term on the right
indicates a summation on σ,
as usual.

And, finally,
writing A_μ' for the two components
represented in $\dfrac{\partial \psi}{\partial x_\mu'}$,

and A_σ for the two components, $\dfrac{\partial \psi}{\partial x_\sigma}$,
we may write (29) as follows:

$$A_\mu' = \frac{\partial x_\sigma}{\partial x_\mu'} \cdot A_\sigma. \qquad (\mu, \sigma = 1, 2) \tag{30}$$

If we now compare (30) with (16)
we note a
VERY IMPORTANT DIFFERENCE,
namely,
that the coefficient on the right in (30)
is the reciprocal of
the coefficient on the right in (16),

171

so that (30) does NOT satisfy
the definition of a vector given in (16).
But it will be remembered that
(16) is the definition of
A CONTRAVARIANT VECTOR ONLY.
And in (30)
we introduce to the reader
the mathematical definition of
A COVARIANT VECTOR.

Note that
to distinguish the two kinds of vectors,
it is customary to write the indices
as SUBscripts in the one case
and as SUPERscripts in the other.*

As before (page 156),
(30) may represent a vector in
any number of dimensions,
depending upon the range of values
given to μ and σ,
and for ANY transformation of coordinates.

Similarly,
A COVARIANT TENSOR OF RANK TWO
is defined by

$$A'_{\alpha\beta} = \frac{\partial x_\gamma}{\partial x'_\alpha} \cdot \frac{\partial x_\delta}{\partial x'_\beta} \cdot A_{\gamma\delta} \qquad (31)$$

and so on, for higher ranks.

COMPARE and CONTRAST carefully
(31) and (17).

*Observe that the SUBscripts are used
 for the COvariant vectors,
 in which the PRIMES in the coefficients
 are in the DENOMINATORS (see (30), p. 171).
 To remember this more easily
 a young student suggests the slogan
 "CO, LOW, PRIMES BELOW."

XIX. MIXED TENSORS.

Addition of covariant vectors
is performed in the same simple manner
as for contravariant vectors (see p. 161).
Thus, the SUM of

$$A'_\lambda = \frac{\partial x_\alpha}{\partial x'_\lambda} \cdot A_\alpha$$

and

$$B'_\lambda = \frac{\partial x_\alpha}{\partial x'_\lambda} \cdot B_\alpha$$

is

$$C'_\lambda = \frac{\partial x_\alpha}{\partial x'_\lambda} \cdot C_\alpha.$$

Also,
the operation defined on page 166
as OUTER MULTIPLICATION
is the same for
covariant tensors:
Thus, the OUTER PRODUCT of

$$A'_\lambda = \frac{\partial x_\alpha}{\partial x'_\lambda} \cdot A_\alpha$$

and

$$B'_{\mu\nu} = \frac{\partial x_\beta}{\partial x'_\mu} \cdot \frac{\partial x_\gamma}{\partial x'_\nu} \cdot B_{\beta\gamma}$$

is

$$C'_{\lambda\mu\nu} = \frac{\partial x_\alpha}{\partial x'_\lambda} \cdot \frac{\partial x_\beta}{\partial x'_\mu} \cdot \frac{\partial x_\gamma}{\partial x'_\nu} \cdot C_{\alpha\beta\gamma}.$$

Furthermore,
it is also possible to multiply
a COVARIANT tensor by
a CONTRAVARIANT one,
thus,

the OUTER PRODUCT of

$$A'_\lambda = \frac{\partial x_\alpha}{\partial x'_\lambda} \cdot A_\alpha$$

and

$$B'^\mu = \frac{\partial x'_\mu}{\partial x_\beta} \cdot B^\beta$$

is

$$C'^\mu_\lambda = \frac{\partial x_\alpha}{\partial x'_\lambda} \cdot \frac{\partial x'_\mu}{\partial x_\beta} \cdot C^\beta_\alpha. \tag{32}$$

Comparison of (32) with (31) and (17)
shows that it is
NEITHER a covariant
NOR a contravariant tensor.
It is called
A MIXED TENSOR of rank TWO.

More generally,
the OUTER PRODUCT of

$$A'^{\alpha\beta}_\gamma = \frac{\partial x_\nu}{\partial x'_\gamma} \cdot \frac{\partial x'_\alpha}{\partial x_\lambda} \cdot \frac{\partial x'_\beta}{\partial x_\mu} \cdot A^{\lambda\mu}_\nu \tag{33}$$

and

$$B'^\kappa_\delta = \frac{\partial x_\sigma}{\partial x'_\delta} \cdot \frac{\partial x'_\kappa}{\partial x_\rho} \cdot B^\rho_\sigma \tag{34}$$

is

$$C'^{\alpha\beta\kappa}_{\gamma\delta} = \frac{\partial x_\nu}{\partial x'_\gamma} \cdot \frac{\partial x_\sigma}{\partial x'_\delta} \cdot \frac{\partial x'_\alpha}{\partial x_\lambda} \cdot \frac{\partial x'_\beta}{\partial x_\mu} \cdot \frac{\partial x'_\kappa}{\partial x_\rho} \cdot C^{\lambda\mu\rho}_{\nu\sigma}. \tag{35}$$

That is,
if any two tensors of ranks m and n,
respectively,
are multiplied together
so as to form their
OUTER PRODUCT,
the result is a TENSOR of rank $m + n$;

thus, the rank of (33) is 3,
and that of (34) is 2,
hence,
the rank of their outer product, (35),
is 5.

Furthermore,
suppose the tensor of rank m
is a MIXED tensor,
having m_1 indices of covariance*
and m_2 indices of contravariance†
(such that $m_1 + m_2 = m$),
and suppose the tensor of rank n
has n_1 indices of covariance*
and n_2 indices of contravariance,†
then,
their outer product will be
a MIXED tensor having
$m_1 + n_1$ indices of covariance*
and
$m_2 + n_2$ indices of contravariance.†

All this has already been illustrated
in the special case given above:
Thus,
(33) has ONE index of covariance (γ)
and (34) also has
ONE index of covariance (δ),
therefore their outer product, (35),
has TWO indices of covariance (γ, δ);
and similarly,
since (33) has
TWO indices of contravariance (α, β)
and (34) has

* SUBscripts.
† SUPERscripts.

ONE index of contravariance (κ),
their outer product, (35),
has
THREE indices of contravariance (α, β, κ).

We hope the reader appreciates
the fact that
although it takes many words
to describe these processes
it is extremely EASY
to DO them
with the AID of the
TENSOR NOTATION.
Thus the outer product of

$$A^{\alpha\beta} \text{ and } B_{\gamma\delta}$$

is simply $C^{\alpha\beta}_{\gamma\delta}$!

Let us remind the reader, however, that
behind this notation,
the processes are really complicated:
Thus (33) represents
a whole SET of equations*
each having MANY* terms on the right.
And (34) also represents
a SET of equations†
each having MANY† terms on the right.
And their outer product, (35),
is obtained by
multiplying

*Namely, EIGHT for two-dimensional space,
 TWENTY-SEVEN for three-dimensional,
 SIXTY-FOUR for four-dimensional,
 and so on.

†Four for two-dimensional space,
 NINE for three-dimensional space,
 SIXTEEN for four-dimensional space,
 and so on.

177

EACH equation of (33) by
EACH one of (34),
resulting in a SET of equations, (35),
containing
THIRTY-TWO equations for
two-dimensional space,
TWO HUNDRED AND FORTY-THREE for
three-dimensional space,
ONE THOUSAND AND TWENTY-FOUR for
four-dimensional space,
and so on.
And all with a
correspondingly large number of terms
on the right of each equation!

And yet
"any child can operate it"
as easily as
pushing a button.

XX. CONTRACTION AND DIFFERENTIATION.

This powerful and
easily operated machine,
the TENSOR CALCULUS,
was devised and perfected by
the mathematicians
Ricci and Levi-Civita
in about 1900,
and was known to very few people
until Einstein made use of it. [12]
Since then it has become
widely known,
and we hope that this little book
will make it intelligible
even to lay readers.

But what use did Einstein make of it?
What is its connection with Relativity?

We are nearly ready to fulfill
the promise made on page 125.

When we have explained
two more operations with tensors,
namely,
CONTRACTION and DIFFERENTIATION,
we shall be able to derive
the promised CURVATURE TENSOR,
from which
Einstein's Law of Gravitation
is obtained.

Consider the mixed tensor (33), p. 175:
suppose we replace in it

$$\gamma \quad \text{by} \quad \alpha,$$

obtaining

$$A'^{\alpha\beta}_{\alpha} = \frac{\partial x_\nu}{\partial x'_\alpha} \cdot \frac{\partial x'_\alpha}{\partial x_\lambda} \cdot \frac{\partial x'_\beta}{\partial x_\mu} \cdot A^{\lambda\mu}_{\nu}. \tag{36}$$

By the summation convention (p. 152),
the left-hand member is to be summed on α,
so that (36) now represents
only TWO equations instead of eight,*
each of which contains
TWO terms on the left instead of one;
furthermore,
on the RIGHT,
since α occurs twice here,
we must sum on α
for each pair of values of ν and λ:
Now,

*See p. 177.

180

when ν happens to have a value
DIFFERENT from λ,

then

$$\frac{\partial x_\nu}{\partial x'_\alpha} \cdot \frac{\partial x'_\alpha}{\partial x_\lambda} = \frac{\partial x_\nu}{\partial x_\lambda} = 0$$

BECAUSE
the x's are NOT functions of each other
(but only of the x''s)
and therefore
there is NO variation of x_ν
with respect to
a DIFFERENT x, namely x_λ.
Thus coefficients of $A_\nu^{\lambda\mu}$ when $\lambda \neq \nu$
will all be ZERO
and will make these terms drop out.
BUT
When $\lambda = \nu$,
then

$$\frac{\partial x_\nu}{\partial x'_\alpha} \cdot \frac{\partial x'_\alpha}{\partial x_\lambda} = \frac{\partial x_\lambda}{\partial x'_\alpha} \cdot \frac{\partial x'_\alpha}{\partial x_\lambda} = 1.$$

Thus (36) becomes

$$A'^{\alpha\beta}_\alpha = \frac{\partial x'_\beta}{\partial x_\mu} \cdot A_\lambda^{\lambda\mu} \tag{37}$$

in which we must still
sum on the right
for λ and μ.

To make all this clearer,
let us write out explicitly
the two equations represented by (37):

$$\begin{cases} A'^{11}_1 + A'^{21}_2 = \dfrac{\partial x'_1}{\partial x_1}\left(A_1^{11} + A_2^{21}\right) + \dfrac{\partial x'_1}{\partial x_2}\left(A_1^{12} + A_2^{22}\right) \\[2ex] A'^{12}_1 + A'^{22}_2 = \dfrac{\partial x'_2}{\partial x_1}\left(A_1^{11} + A_2^{21}\right) + \dfrac{\partial x'_2}{\partial x_2}\left(A_1^{12} + A_2^{22}\right). \end{cases}$$

Thus (37) may be written more briefly:

$$C'^{\beta} = \frac{\partial x'_{\beta}}{\partial x_{\mu}} \cdot C^{\mu} \qquad (38)$$

where

$$C'^{1} = A'^{11}_{1} + A'^{21}_{2},$$
$$C'^{2} = A'^{12}_{1} + A'^{22}_{2}$$

and

$$C^{1} = A^{11}_{1} + A^{21}_{2},$$
$$C^{2} = A^{12}_{1} + A^{22}_{2}.$$

In other words,
by making one upper and one lower index
ALIKE
in (33),
we have REDUCED
a tensor of rank THREE to
a tensor of rank ONE.

The important thing to note is
that this process of reduction
or CONTRACTION,
as it is called,
leads again to
A TENSOR,
and it is obvious that
for every such contraction
the rank is reduced by TWO,
since for every such contraction
two of the partial derivatives in
the coefficient
cancel out (see page 181).

We shall see later
how important this process of contraction is.

Now,
if we form the OUTER PRODUCT of two tensors,
in the way already described (p. 175)

182

and if the result is
a mixed tensor,
then,
by contracting this mixed tensor
as shown above,
we get a tensor which is called
an INNER PRODUCT
in contrast to
their OUTER PRODUCT.

Thus the OUTER product of

$$A_{\alpha\beta} \quad \text{and} \quad B^{\gamma}$$

is

$$C^{\gamma}_{\alpha\beta}$$

(see page 177);
now, if in this result
we replace γ by β,
obtaining

$$C^{\beta}_{\alpha\beta} \text{ or } D_{\alpha} \text{ (see pages 180 to 182)},$$

then $\qquad D_{\alpha}$ is an INNER product of

$A_{\alpha\beta}$ and B^{γ}.

And now we come to
DIFFERENTIATION.

We must remind the reader that
if

$$y = uv$$

where y, u, and v are variables,
then*

$$\frac{dy}{dx} = u\frac{dv}{dx} + v\frac{du}{dx}.$$

Applying this principle to
the differentiation of

$$A'^{\mu} = \frac{\partial x'_{\mu}}{\partial x_{\sigma}} \cdot A^{\sigma}, \tag{39}$$

*See any calculus textbook, e. g. [9].

183

with respect to x'_ν,
we get:

$$\frac{\partial A'^\mu}{\partial x'_\nu} = \frac{\partial x'_\mu}{\partial x_\sigma} \cdot \frac{\partial A^\sigma}{\partial x'_\nu} + A^\sigma \cdot \frac{\partial^2 x'_\mu}{\partial x_\sigma \cdot \partial x'_\nu}. \qquad (40)$$

Or, since

$$\frac{\partial A^\sigma}{\partial x'_\nu} = \frac{\partial A^\sigma}{\partial x_\tau} \cdot \frac{\partial x_\tau}{\partial x'_\nu}, \quad \text{by (26),}$$

hence (40) becomes

$$\frac{\partial A'^\mu}{\partial x_\nu} = \frac{\partial x'_\mu}{\partial x_\sigma} \cdot \frac{\partial x_\tau}{\partial x'_\nu} \cdot \frac{\partial A^\sigma}{\partial x_\tau} + \frac{\partial^2 x'_\mu}{\partial x_\sigma \cdot \partial x'_\nu} \cdot A^\sigma. \qquad (41)$$

From (41) we see that
if the second term on the right
were not present,
then (41) would represent
a mixed tensor of rank two.
And, in certain special cases,
this second term does vanish,
so that
in SUCH cases,
differentiation of a tensor
leads to another tensor
whose rank is one more than
the rank of the given tensor.
Such a special case is the one
in which the coefficients

$$\frac{\partial x'_\mu}{\partial x_\sigma}$$

in (39)
are constants,
as in (13) on page 147,
since the coefficients in (13)
are the same as those in (11) or (10);

and are therefore functions of θ,
θ being the angle through which
the axes were rotated (page 141),
and therefore a constant.
In other words,
when the transformation of coordinates
is of the simple type
described on page 141,
then
ordinary differentiation of a tensor
leads to a tensor.

BUT, IN GENERAL,
these coefficients are NOT constants,
and so,
IN GENERAL
differentiation of a tensor
does NOT give a tensor
as is evident from (41).

BUT
there is a process called
COVARIANT DIFFERENTIATION
which ALWAYS leads to a tensor,
and which we shall presently describe.

We cannot emphasize too often
the IMPORTANCE
of any process which
leads to a tensor,
since tensors represent
the "FACTS" of our universe
(see page 149).

And, besides,
we shall have to employ
COVARIANT DIFFERENTIATION
in deriving

the long-promised
CURVATURE TENSOR
and
EINSTEIN'S LAW OF GRAVITATION.

XXI. THE LITTLE g's.

To explain covariant differentiation
we must first refer the reader back
to chapter XIII,
in which it was shown that
the distance between two points,
or, rather, the square of this distance,
namely, ds^2,
takes on various forms
depending upon
(a) the surface in question
and
(b) the coordinate system used.

But now,
with the aid of the remarkable notation
which we have since explained,
we can include
ALL these expressions for ds^2
in the SINGLE expression

$$ds^2 = g_{\mu\nu} \cdot dx_\mu \cdot dx_\nu; \qquad (42)$$

and, indeed,
this holds NOT ONLY for
ANY SURFACE,
but also for
any THREE-dimensional space,
or FOUR-dimensional,

or, in general,
any *n*-dimensional space!*

Thus, to show how (42) represents
equation (3) on page 116,
we take $\mu = 1, 2$ and $\nu = 1, 2$,
obtaining

$$
\begin{aligned}
ds^2 = \; & g_{11}dx_1 \cdot dx_1 + g_{12}dx_1 \cdot dx_2 \\
& + g_{21}dx_2 \cdot dx_1 + g_{22}dx_2 \cdot dx_2,
\end{aligned}
\tag{43}
$$

since the presence of μ and ν
TWICE
in the term on the right in (42)
requires SUMMATION on both μ and ν.†
Of course (43) may be written:‡

$$
\begin{aligned}
ds^2 = \; & g_{11}dx_1^2 + g_{12}dx_1dx_2 \\
& + g_{21}dx_2dx_1 + g_{22}dx_2^2;
\end{aligned}
\tag{44}
$$

and, comparing (44) with (3),
we find that
the coefficients in (3)
have the particular values:

$$
g_{11} = 1, \quad g_{12} = 0, \quad g_{21} = 0, \quad g_{22} = 1.
$$

*Except only at a so-called "singular point"
 of a space;
 that is,
 a point at which
 matter is actually located.
 In other words,
 (42) holds for any region AROUND matter.

†See page 152.

‡Note that in dx_1^2 (as well as in dx_2^2)
 the upper "2" is really an exponent
 and NOT a SUPERSCRIPT
 since (44) is an
 ordinary algebraic equation
 and is NOT in the
 ABBREVIATED TENSOR NOTATION.

188

Similarly, in (6) on page 120,

$$g_{11} = r^2, \quad g_{12} = 0, \quad g_{21} = 0, \quad g_{22} = R^2;$$

and, in (7) on page 123,

$$g_{11} = 1, \quad g_{12} = -\cos\alpha, \quad g_{21} = -\cos\alpha, \quad g_{22} = 1,$$

and so on.

Note that g_{12} and g_{21} have
the SAME value.
And indeed, in general

$$g_{\mu\nu} = g_{\nu\mu}$$

in (42) on page 187.

Of course, if, in (42),
we take $\mu = 1, 2, 3$ and $\nu = 1, 2, 3$,
we shall get the value for ds^2
in a THREE-dimensional space:

$$\begin{aligned}
ds^2 = {} & g_{11}dx_1^2 + g_{12}dx_1 \cdot dx_2 + g_{13}dx_1 \cdot dx_3 \\
& + g_{21}dx_2 \cdot dx_1 + g_{22}dx_2^2 + g_{23}dx_2 \cdot dx_3 \qquad (45) \\
& + g_{31}dx_3 \cdot dx_1 + g_{32}dx_3 \cdot dx_2 + g_{33}dx_3^2.
\end{aligned}$$

Thus, in particular,
for ordinary Euclidean three-space,
using the common rectangular coordinate,
we now have:

$$g_{11} = 1, \quad g_{22} = 1, \quad g_{33} = 1,$$

and all the other g's are zero,
so that (42) becomes,
for THIS PARTICULAR CASE,
the familiar expression

$$ds^2 = dx_1^2 + dx_2^2 + dx_3^2$$

or

$$ds^2 = dx^2 + dy^2 + dz^2;$$

and similarly for
higher dimensions.

Thus, for a given space,
two-, three-, four-, or n-dimensional,
and for a given set of coordinates,
we get a certain set of g's.

It is easy to show* that
any such set of g's,
(which is represented by $g_{\mu\nu}$)
constitutes
the COMPONENTS of a TENSOR,
and, in fact, that
it is a
COVARIANT TENSOR OF RANK TWO,
and hence is appropriately
designated with TWO SUBscripts†:

$$g_{\mu\nu}.$$

Let us now briefly sum up
the story so far:

By introducing
the Principle of Equivalence
Einstein replaced the idea of
a "force of gravity"
by the concept of
a geometrical space (Chap. XII).
And since a space
is characterized by its g's,
the knowledge of the g's of a space
is essential to a study of
how things move in the space,
and hence essential
to an understanding of
Einstein's Law of Gravitation.

*See page 313.
†See page 172.

190

XXII. OUR LAST DETOUR.

As we said before (page 185),
to derive the
Einstein Law of Gravitation,
we must employ
COVARIANT DIFFERENTIATION.
Now, the COVARIANT DERIVATIVE of a tensor
contains certain quantities known as
CHRISTOFFEL SYMBOLS*
which are functions of the tensor $g_{\mu\nu}$
discussed in chapter XXI,
and also of another set $g^{\mu\nu}$
(note the SUPERscripts here)
which we shall now describe: [13]

For simplicity,
let us limit ourselves for the moment
to TWO-dimensional space,
that is,
let us take $\mu = 1, 2$ and $\nu = 1, 2$;
then $g_{\mu\nu}$ will have
FOUR components,
namely,
the four coefficients on the right
in (44).
And let us arrange these coefficients
in a SQUARE ARRAY, thus:

$$\left\| \begin{matrix} g_{11} & g_{12} \\ g_{21} & g_{22} \end{matrix} \right\|$$

which is called a MATRIX.
Now since $g_{12} = g_{21}$ (see page 189)

*Introduced by the mathematician E. B. Christoffel (1869).

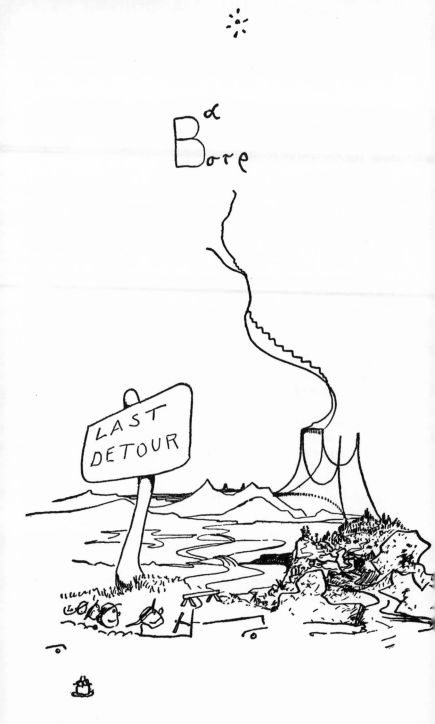

this is called a
SYMMETRIC MATRIX,
since it is symmetric with respect to
the principal diagonal
(that is, the one which starts
in the upper left-hand corner).

If we now replace the double bars
on each side of the matrix
by SINGLE bars,
as shown in the following:

$$\begin{vmatrix} g_{11} & g_{12} \\ g_{21} & g_{22} \end{vmatrix}$$

we get what is known as
a DETERMINANT.*
The reader must carefully
DISTINGUISH between

*The reader probably knows that
a square array of numbers
with single bars on each side

$$\begin{vmatrix} 5 & 6 \\ 2 & 3 \end{vmatrix}$$

is called a determinant,
and that its value is found thus:

$$5 \times 3 - 6 \times 2 = 15 - 12 = 3.$$

Or, more generally,

$$\begin{vmatrix} a & b \\ c & d \end{vmatrix} = ad - bc.$$

A determinant does not necessarily
have to have TWO rows and columns,
but may have n rows and n columns,
and is then said to be of order n.
The way to find the VALUE of
a determinant of the nth order
is described in many
college algebra textbooks. [10]

193

a square array with SINGLE bars
from one with DOUBLE bars:
the FORMER is a DETERMINANT
and has a SINGLE VALUE
found by combing the "elements"
in a certain way
as mentioned in the footnote on p. 193.
Whereas,
the DOUBLE-barred array
is a set of SEPARATE "elements,"
NOT to be COMBINED in any way.
They may be just
the coefficients of the separate terms
on the right in (44),
which,
as we mentioned on page 190,
are the separate COMPONENTS of a tensor.

The determinant on page 193
may be designated more briefly by

$$ \mid g_{\mu\nu} \mid, \qquad (\mu = 1, 2;\ \nu = 1, 2) $$

or, still better, simply by g.

And now let us form a new square array
in the following manner:
DIVIDE the COFACTOR* of EACH ELEMENT
of the determinant on page 193
by the value of the whole determinant,
namely, by g,
thus obtaining the corresponding element of
the NEW array.

*For readers unfamiliar with determinants
this term is explained on p. 195.

The COFACTOR of a given element
of a determinant
is found by striking out
the row and column containing the given element,
and evaluating the
determinant which is left over,
prefixing the sign $+$ or $-$
according to a certain rule:
Thus, in the determinant

$$\begin{vmatrix} 5 & 2 & 3 \\ 4 & 1 & 0 \\ 6 & 8 & 7 \end{vmatrix}$$

the cofactor of the element 5, is:

$$+ \begin{vmatrix} 1 & 0 \\ 8 & 7 \end{vmatrix} = 1 \times 7 - 8 \times 0 = 7 - 0 = 7.$$

Similarly, the cofactor of 4 is

$$- \begin{vmatrix} 2 & 3 \\ 8 & 7 \end{vmatrix} = -(14 - 24) = 10,$$

and so on.
Note that in the first case
we prefixed the sign $+$,
while in the second case
we prefixed a $-$.
The rule is:
prefix a $+$ or $-$ according as
the NUMBER of steps required to go
from the first element
(that is, the one in
the upper left-hand corner)
to the given element,
is EVEN or ODD, respectively;
thus to go from "5" to "4"
it takes one step,
hence the cofactor of "4" must have
a MINUS prefixed before

$$\begin{vmatrix} 2 & 3 \\ 8 & 7 \end{vmatrix}.$$

But all this is more thoroughly
explained in most college textbooks
on algebra and trigonometry. [10]

195

Let us now go back
to the array described
at the bottom of p. 194.

This new array,
which we shall designate by $g^{\mu\nu}$
can also be shown to be
a TENSOR,
and, this time,
A CONTRAVARIANT TENSOR
OF RANK TWO.
That it is also SYMMETRIC
can easily be shown by the reader.

We can now give the definition
of the Christoffel symbol
which we need.
It is designated by $\{\mu\nu, \lambda\}$
and is a symbol for:

$$\tfrac{1}{2}g^{\lambda\alpha}\left(\frac{\partial g_{\mu\alpha}}{\partial x_\nu} + \frac{\partial g_{\nu\alpha}}{\partial x_\mu} - \frac{\partial g_{\mu\nu}}{\partial x_\alpha}\right). \tag{46}$$

In other words,
the above-mentioned Christoffel symbol*
involves partial derivatives of
the coefficients in (44),
combined as shown in (46)
and multiplied by
the components of the tensor $g^{\mu\nu}$.
Thus, in two-dimensional space,

*There are other Christoffel symbols, [14]
but we promised the reader
to introduce only the
barest minimum of mathematics
necessary for our purpose!

since μ, ν, λ, α each have the values 1, 2,
we have, for example,

$$\{11, 1\} = \tfrac{1}{2}g^{11}\left(\frac{\partial g_{11}}{\partial x_1} + \frac{\partial g_{11}}{\partial x_1} - \frac{\partial g_{11}}{\partial x_1}\right)$$
$$+ \tfrac{1}{2}g^{12}\left(\frac{\partial g_{12}}{\partial x_1} + \frac{\partial g_{12}}{\partial x_1} - \frac{\partial g_{11}}{\partial x_2}\right),$$

and similarly for the remaining
SEVEN values of

$$\{\mu\nu, \lambda\}$$

obtained by allowing μ, ν, and λ
to take on their two values for each.

Note that in evaluating $\{11, 1\}$ above,
we SUMMED on the α,
allowing α to take on BOTH values, 1, 2,
BECAUSE
if (46) were multiplied out,
EACH TERM would contain α TWICE,
and this calls for
SUMMATION on the α (see page 152).
Now that we know the meaning of
the 3-index Christoffel symbol

$$\{\mu\nu, \lambda\},$$

we are ready to define
the covariant derivative of a tensor,
from which it is only a step to
the new Law of Gravitation.

If A_σ is a covariant tensor of rank one,
its COVARIANT DERIVATIVE
with respect to x_τ
is DEFINED as:

$$\frac{\partial A_\sigma}{\partial x_\tau} - \{\sigma\tau, \alpha\}\, A_\alpha. \tag{47}$$

It can be shown to be a TENSOR —
in fact, it is a
COVARIANT TENSOR OF RANK TWO*
and may therefore be designated by

$$A_{\sigma\tau}.$$

Similarly,
if we have
a contravariant tensor of rank one,
represented by A^σ,
its COVARIANT DERIVATIVE
with respect to x_τ
is the TENSOR:

$$A^\sigma_\tau = \frac{\partial A^\sigma}{\partial x_\tau} + \{\tau\epsilon, \sigma\} A^\epsilon. \qquad (48)$$

Or,
starting with tensors of rank TWO,
we have the following three cases:

(a) starting with the
CONTRAVARIANT tensor, $A^{\sigma\tau}$,
we get the COVARIANT DERIVATIVE:

$$A^{\sigma\tau}_\rho = \frac{\partial A^{\sigma\tau}}{\partial x_\rho} + \{\rho\epsilon, \sigma\} A^{\epsilon\tau} + \{\rho\epsilon, \tau\} A^{\sigma\epsilon},$$

(b) from the MIXED tensor, A^τ_σ,
we get the COVARIANT DERIVATIVE:

$$A^\tau_{\sigma\rho} = \frac{\partial A^\tau_\sigma}{\partial x_\rho} - \{\rho\sigma, \epsilon\} A^\tau_\epsilon + \{\rho\epsilon, \tau\} A^\epsilon_\sigma,$$

*See Eddington, *Mathematical Theory of Relativity*, pp. 60–61, 65–66;
or d'Inverno, *Introducing Einstein's Relativity*, pp. 72–74,
in Further Reading.

(c) from the COVARIANT tensor, $A_{\sigma\tau}$,
we get the COVARIANT DERIVATIVE:

$$A_{\sigma\tau\rho} = \frac{\partial A_{\sigma\tau}}{\partial x_\rho} - \{\sigma\rho, \epsilon\} A_{\epsilon\tau} - \{\tau\rho, \epsilon\} A_{\sigma\epsilon}.$$

And similarly for the
COVARIANT DERIVATIVES
of tensors of higher ranks.

Note that IN ALL CASES
COVARIANT DIFFERENTIATION
OF A TENSOR
leads to a TENSOR having
ONE MORE UNIT OF
COVARIANT CHARACTER
than the given tensor.

Of course since
the covariant derivative of a tensor
is itself a tensor,
we may find
ITS covariant derivative
which is then the
SECOND COVARIANT DERIVATIVE of
the original tensor,
and so on for
higher covariant derivatives.

Note also that
when the g's happen to be constants,
as, for example,
in the case of a Euclidean plane,
using rectangular coordinates,
in which case we have (see p. 188)

$$ds^2 = dx_1^2 + dx_2^2,$$

so that

$$g_{11} = 1,\ g_{12} = 0,\ g_{21} = 0,\ g_{22} = 1,$$

199

all constants,
then obviously
the Christoffel symbols here
are all ZERO,
since the derivative of a constant is zero,
and every term of the
Christoffel symbol
has such a derivative as a factor,*

so that (47) becomes simply $\dfrac{\partial A_\sigma}{\partial x_\tau}$.

That is, in this case,
the covariant derivative becomes
simply the ordinary derivative.
But of course
this is NOT so IN GENERAL.

XXIII. THE CURVATURE TENSOR AT LAST.

Having now built up the necessary machinery,
the reader will have no trouble
in following the derivation of
the new Law of Gravitation.

Starting with the tensor, A_σ,
form its covariant derivative (see p. 197)
with respect to x_τ:

$$A_{\sigma\tau} = \frac{\partial A_\sigma}{\partial x_\tau} - \{\sigma\tau, \alpha\}\, A_\alpha. \qquad (49)$$

*See page 196.

Now form the covariant derivative
of $A_{\sigma\tau}$ (see page 199)
with respect to x_ρ:

$$A_{\sigma\tau\rho} = \frac{\partial A_{\sigma\tau}}{\partial x_\rho} - \{\sigma\rho, \epsilon\}\, A_{\epsilon\tau} - \{\tau\rho, \epsilon\}\, A_{\sigma\epsilon} \qquad (50)$$

obtaining
a SECOND covariant derivative of A_σ,
which is a
COVARIANT TENSOR
OF RANK THREE.
Substituting (49) in (50),
we get

$$A_{\sigma\tau\rho} = \frac{\partial^2 A_\sigma}{\partial x_\tau \partial x_\rho} - \{\sigma\tau, \alpha\} \frac{\partial A_\alpha}{\partial x_\rho} - A_\alpha \frac{\partial}{\partial x_\rho} \{\sigma\tau, \alpha\}$$

$$- \{\sigma\rho, \epsilon\} \left[\frac{\partial A_\epsilon}{\partial x_\tau} - \{\epsilon\tau, \alpha\}\, A_\alpha \right]$$

$$- \{\tau\rho, \epsilon\} \left[\frac{\partial A_\sigma}{\partial x_\epsilon} - \{\sigma\epsilon, \alpha\}\, A_\alpha \right]$$

or

$$A_{\sigma\tau\rho} = \frac{\partial^2 A_\sigma}{\partial x_\tau \partial x_\rho} - \{\sigma\tau, \alpha\} \frac{\partial A_\alpha}{\partial x_\rho} - A_\alpha \frac{\partial}{\partial x_\rho} \{\sigma\tau, \alpha\}$$

$$- \{\sigma\rho, \epsilon\} \frac{\partial A_\epsilon}{\partial x_\tau} + \{\sigma\rho, \epsilon\} \{\epsilon\tau, \alpha\}\, A_\alpha \qquad (51)$$

$$- \{\tau\rho, \epsilon\} \frac{\partial A_\sigma}{\partial x_\epsilon} + \{\tau\rho, \epsilon\} \{\sigma\epsilon, \alpha\}\, A_\alpha.$$

If we had taken these derivatives
in the REVERSE order,
namely,
FIRST with respect to x_ρ
and THEN with respect to x_τ,
we would of course have obtained
the following result instead:

$$A_{\sigma\rho\tau} = \frac{\partial^2 A_\sigma}{\partial x_\rho \partial x_\tau} - \{\sigma\rho, \alpha\} \frac{\partial A_\alpha}{\partial x_\tau} - A_\alpha \frac{\partial}{\partial x_\tau} \{\sigma\rho, \alpha\}$$

$$- \{\sigma\tau, \epsilon\} \frac{\partial A_\epsilon}{\partial x_\rho} + \{\sigma\tau, \epsilon\} \{\epsilon\rho, \alpha\} A_\alpha \qquad (52)$$

$$- \{\rho\tau, \epsilon\} \frac{\partial A_\sigma}{\partial x_\epsilon} + \{\rho\tau, \epsilon\} \{\sigma\epsilon, \alpha\} A_\alpha$$

which is again
a COVARIANT TENSOR OF RANK THREE.

Now,
comparing (51) with (52)
we shall find that they are
NOT alike THROUGHOUT:
Only SOME of the terms are the
SAME in both,
but the remaining terms are different.

Let us see:

the FIRST term in each is:

$$\frac{\partial^2 A_\sigma}{\partial x_\tau \partial x_\rho} \quad \text{and} \quad \frac{\partial^2 A_\sigma}{\partial x_\rho \partial x_\tau}, \quad \text{respectively.}$$

These, by ordinary calculus,

ARE the same.*

The SECOND term of (51)

is the same as

the FOURTH term of (52)

since the occurrence of α (or ϵ)

TWICE in the same term

implies a SUMMATION

and it is therefore immaterial

what letter is used (α or ϵ)! †

Similarly for

the FOURTH term of (51)

and the SECOND of (52).

The SIXTH term and the SEVENTH

are the same in both

since the reversal of τ and ρ in

$$\{\tau\rho, \epsilon\}$$

---

*For, suppose that z is a function of x and y, as, for example, $z = x^2 + 2xy$.

Then $\dfrac{\partial z}{\partial x} = 2x + 2y$ (treating y as constant)

and $\dfrac{\partial^2 z}{\partial y \cdot \partial x} = 2$ (treating x as constant).

And, if we reverse the order of differentiation, finding FIRST the derivative with respect to y and THEN with respect to x, we would get

$\dfrac{\partial z}{\partial y} = 2x$ (treating x as constant)

and $\dfrac{\partial^2 z}{\partial x \cdot \partial y} = 2$ (treating y as constant).

the SAME FINAL result.
And this is true IN GENERAL.

†An index which is thus easily replaceable is called a "dummy"!

does not alter the value of
the Christoffel symbol:
This can easily be seen by referring
to the definition of this symbol,*
and remembering that the tensor $g_{\mu\nu}$
is SYMMETRIC,

that is, $g_{\mu\nu} = g_{\nu\mu}$ (see page 189).

Similarly the last term is the same
in both (51) and (52).

But the THIRD and FIFTH terms of (51)
are NOT equal to any of the terms in (52).
Hence by subtraction we get

$$A_{\sigma\tau\rho} - A_{\sigma\rho\tau} = \{\sigma\rho, \epsilon\}\{\epsilon\tau, \alpha\}A_\alpha - A_\alpha \frac{\partial}{\partial x_\rho}\{\sigma\tau, \alpha\}$$

$$+ A_\alpha \frac{\partial}{\partial x_\tau}\{\sigma\rho, \alpha\} - \{\sigma\tau, \epsilon\}\{\epsilon\rho, \alpha\}A_\alpha$$

or

$$A_{\sigma\tau\rho} - A_{\sigma\rho\tau} = \left[\frac{\partial}{\partial x_\tau}\{\sigma\rho, \alpha\} - \frac{\partial}{\partial x_\rho}\{\sigma\tau, \alpha\} \right.$$

$$\left. + \{\sigma\rho, \epsilon\}\{\epsilon\tau, \alpha\} - \{\sigma\tau, \epsilon\}\{\epsilon\rho, \alpha\} \right] A_\alpha. \tag{53}$$

And since addition (or subtraction)
of tensors
gives a result which is itself a tensor (see page 161)
the left-hand member of (53) is
A COVARIANT TENSOR OF RANK THREE,
hence of course the right-hand member
is also such a tensor.
But, now,
since A_α is an arbitrary covariant vector,

*See page 196.

205

its coefficient,
namely, the quantity in square brackets,
must also be a tensor
according to the theorem on p. 312.
Furthermore,
this bracketed expression
must be a MIXED tensor of RANK FOUR,
since on inner multiplication by A_α
it must give a result which is
of rank THREE;
and indeed it must be of the form [15]

$$B^\alpha_{\sigma\tau\rho}$$

(see page 313).
This
AT LAST
is the long-promised
CURVATURE TENSOR (page 187),
and is known as
THE RIEMANN-CHRISTOFFEL TENSOR.

Let us examine it carefully
so that we may appreciate
its meaning and value.

XXIV. OF WHAT USE IS THE CURVATURE TENSOR?

In the first place
we must remember that
it is an abbreviated notation for
the expression in square brackets
in (53) on page 205;
in which,
if we substitute for the Christoffel symbols,

$\{\sigma\rho, \epsilon\}$ and so on,
their values in accordance with
the definition on page 196,
we find that we have
an expression containing
first and second partial derivatives
of the g's,
which are themselves the coefficients
in the expression for ds^2 (see p. 187).

How many components does
the Riemann-Christoffel tensor have?
Obviously that depends upon the
dimensionality of the space
under consideration.
Thus, if we are studying
a two-dimensional surface,
then each of the indices
will have two possible values,
so that $B^{\alpha}_{\sigma\tau\rho}$ would then have
sixteen components.
Similarly,
in three-dimensional space
it would have 3^4 or 81 components,
and so on.
For the purposes of Relativity,
in which we have to deal with
a FOUR-dimensional continuum
this tensor has 4^4 or 256 components!

We hasten to add that
it is not quite so bad as that,
as we can easily see:
In the first place,
if, in this tensor,*

*That is, in the expression in square brackets
in (53) on page 205.

we interchange τ and ρ,
the result is merely to change its sign.*
Hence,
of the possible 16 combinations of τ and ρ,
only 6 are independent:
This is in itself so interesting
that we shall linger here for a moment:
Suppose we have 16 quantities, $a_{\alpha\beta}$,
(where $\alpha = 1, 2, 3, 4$, and $\beta = 1, 2, 3, 4$),
which we may arrange as follows:

$$\left\| \begin{array}{cccc} a_{11} & a_{12} & a_{13} & a_{14} \\ a_{21} & a_{22} & a_{23} & a_{24} \\ a_{31} & a_{32} & a_{33} & a_{34} \\ a_{41} & a_{42} & a_{43} & a_{44} \end{array} \right\|$$

And suppose that $a_{\alpha\beta} = -a_{\beta\alpha}$
(that is, a reversal of the two subscripts
results only in a change of sign of the term),
then, since $a_{11} = -a_{11}$ implies that $a_{11} = 0$,
and similarly for the remaining terms
in the principal diagonal,
hence,
the above array becomes:

$$\left\| \begin{array}{cccc} 0 & a_{12} & a_{13} & a_{14} \\ -a_{12} & 0 & a_{23} & a_{24} \\ -a_{13} & -a_{23} & 0 & a_{34} \\ -a_{14} & -a_{24} & -a_{34} & 0 \end{array} \right\|$$

Thus there are only
SIX distinct quantities
instead of sixteen.
Such an array is called
ANTISYMMETRIC.

*The reader would do well to compare this expression with the one obtained from it by an interchange of τ and ρ throughout.

Compare this with the definition of
a SYMMETRIC array on page 193.*
And so,
to come back to the discussion on page 208,
we now have
six combinations of τ and ρ
to be used with
sixteen combinations of σ and α,
giving 6×16 or 96 components
instead of 256.

Furthermore,
it can be shown†
that we can further reduce this number
to 20.
Thus our curvature tensor,
for the situation in Relativity,
has only 20 components and NOT 256!

Now let us consider for a moment
the great IMPORTANCE of this tensor
in the study of spaces.

*Thus in an ANTISYMMETRIC matrix we have

$$a_{\alpha\beta} = -a_{\beta\alpha},$$

whereas, in a SYMMETRIC matrix we have

$$a_{\alpha\beta} = a_{\beta\alpha}.$$

Note that if the first matrix on p. 208
were SYMMETRIC,
it would reduce to
TEN distinct elements,
since the elements in the principal diagonal
would NOT be zero in that case.

†See Eddington, p. 72,
or d'Inverno, p. 86,
in Further Reading.

Suppose we have
a Euclidean space
of two, three, or more dimensions,
and suppose we use
ordinary rectangular coordinates.
Here the g's are all constants.*
Hence,
since the derivative of a constant
is zero
the Christoffel symbols will
also be zero (see page 200);
and, therefore,
all the components of the
curvature tensor
will be zero too,
because every term contains
a Christoffel symbol (see page 205).

BUT,
if the components of a tensor
in any given coordinate system
are all zero,
obviously its components in
any other coordinate system
would also be zero
(consider this in the simple case on page 129).

And so,
whereas from a mere superficial inspection
of the expression for ds^2
we cannot tell whether
the space is Euclidean or not,†
an examination of the curvature tensor
(which of course is obtained
from the coefficients
in the expression for ds^2)

*See page 189.

†See page 125.

can definitely give this information,
no matter what coordinate system
is used in setting up ds^2.
Thus,
whether we use (3) on page 116
or (7) or (8) on page 123,
all of which represent
the square of the distance
between two points
ON A EUCLIDEAN PLANE,
using various coordinate systems,
we shall find that
the components of $B^\alpha_{\sigma\tau\rho}$ in all three cases
ARE ALL ZERO.*
The same is true
for all coordinate systems
and for any number of dimensions,
provided that we remain in
Euclidean geometry.

---

*To have a clear idea of
the meaning of the symbolism,
the reader should try the simple exercise
of showing that $B^\alpha_{\sigma\tau\rho} = 0$ for (8) on p. 123.
She must bear in mind that here

$$g_{11} = 1, \quad g_{12} = g_{21} = 0, \quad g_{22} = x_1^2,$$

and use these values in the bracketed expression
in (53) on page 205,
remembering of course that the meaning of $\{\sigma\rho, \epsilon\}$, etc.
is given by the definition on page 196;
also that all indices, σ, ρ, ϵ, etc.
have the possible values 1 and 2,
since the space here is
two-dimensional;
and she must not forget to SUM
whenever an index appears
TWICE IN ANY ONE TERM.

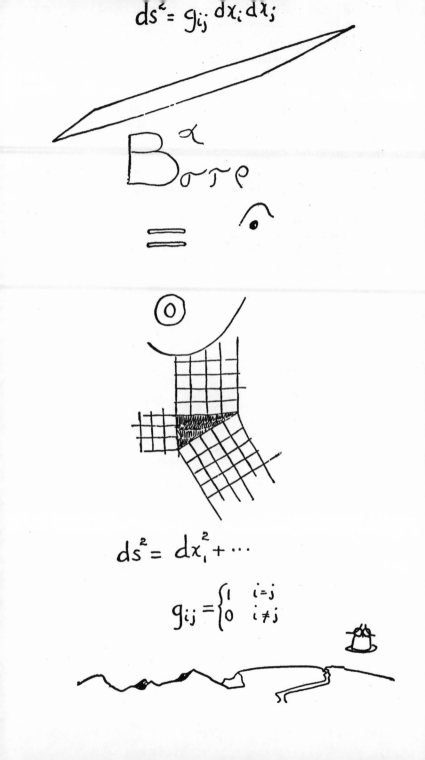

Thus

$$B^{\alpha}_{\sigma\tau\rho} = 0 \qquad (54)$$

is a NECESSARY condition
that a space shall be
EUCLIDEAN.

It can be shown that
this is also a SUFFICIENT condition.

In other words,
given a Euclidean space,
this tensor will be zero,
whatever coordinate system is used,
AND CONVERSELY,
given this tensor equal to zero,
then we know that
the space must be Euclidean.

We shall now see
how the new Law of Gravitation
is EASILY derived
from this tensor.

XXV. THE BIG G'S OR EINSTEIN'S LAW OF GRAVITATION.

In (54) replace ρ by α,
obtaining

$$B^{\alpha}_{\sigma\tau\alpha} = 0. \qquad (55)$$

Since α appears twice
in the term on the left,
we must,
according to the usual convention,
sum on α,

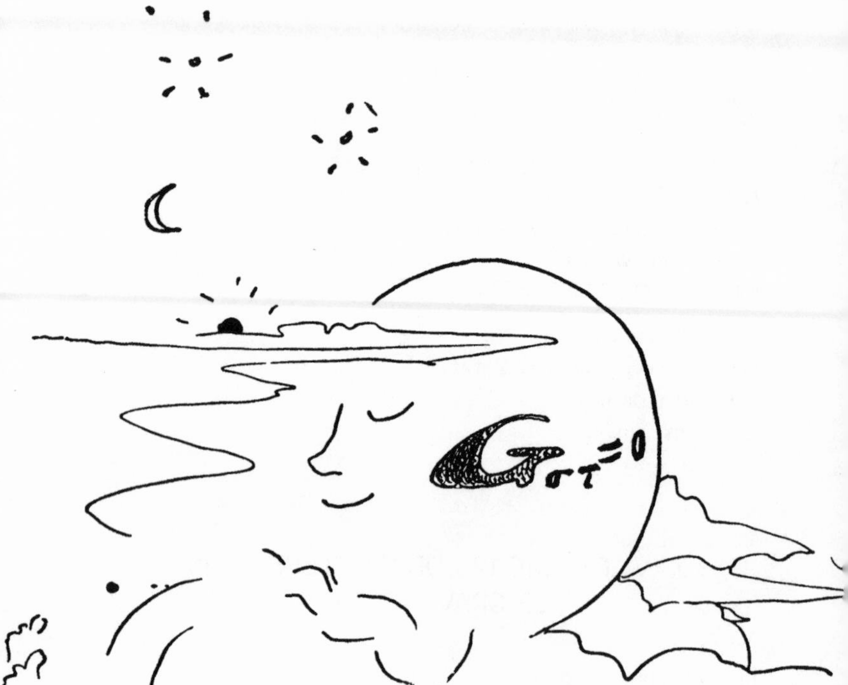
$G_{\sigma\tau}=0$

so that (55) represents
only SIXTEEN equations
corresponding to the
4 × 4 values of σ and τ
in a four-dimensional continuum.
Thus, when $\sigma = \tau = 1$,
(55) becomes

$$B^1_{111} + B^2_{112} + B^3_{113} + B^4_{114} = 0.$$

Similarly, for $\sigma = 1$, $\tau = 2$,
we get

$$B^1_{121} + B^2_{122} + B^3_{123} + B^4_{124} = 0$$

and so on,
for the 16 possible combinations of σ and τ.
We may therefore write (55) in the form [16]

$$G_{\sigma\tau} = 0 \qquad\qquad (56)$$

where each G consists of 4 B's
as shown above.
In other words,
by CONTRACTING $B^\alpha_{\sigma\tau\rho}$,
which is a tensor of the FOURTH rank,
we get a tensor of the SECOND rank,
namely, $G_{\sigma\tau}$,
as explained on page 182.

The QUITE INNOCENT-LOOKING
EQUATION (56) IS
EINSTEIN'S LAW OF GRAVITATION.

Perhaps the reader is startled
by this sudden announcement.
But let us look into (56)
carefully,
and see what is behind its
innocent simplicity,

and why it deserves to be called
the Law of Gravitation.

In the first place
it must be remembered
that before contraction,

$$B^{\alpha}_{\sigma\tau\rho}$$

represented the quantity in brackets
in the right-hand member
of equation (53) on page 205.
Hence,
when we contracted it
by replacing ρ by α,
we can see from (53) that
$G_{\sigma\tau}$ represents
the following expression:

$$\frac{\partial}{\partial x_\tau}\{\sigma\alpha,\alpha\} - \frac{\partial}{\partial x_\alpha}\{\sigma\tau,\alpha\}$$
$$+ \{\sigma\alpha,\epsilon\}\{\epsilon\tau,\alpha\} - \{\sigma\tau,\epsilon\}\{\epsilon\alpha,\alpha\}, \tag{57}$$

which, in turn,
by the definition of
the Christoffel symbol (page 196)
represents
an expression containing
first and second partial derivatives
of the little g's.
And, of course,
(57) takes 16 different values
as σ and τ each take on
their 4 different values,
while the other Greek letters in (57),
namely, α and ϵ,
are mere dummies (see page 204)
and are to be summed
(since each occurs twice in each term),
as usual.

To get clearly in mind
just what (57) means,
the reader is advised
to replace each Christoffel symbol
in accordance with the definition on page 196,
and to write out in particular
one of the 16 expressions represented by (57)
by putting, say, $\sigma = 1$ and $\tau = 2$,
and allowing α and ϵ to assume,
in succession,
the values 1, 2, 3, 4.

It can easily be shown
that (56) actually represents
NOT 16 DIFFERENT equations
but only 10,
and, of these, only 6 are independent.*
So that the new Law of Gravitation
is not quite so complicated
as it appears at first.

But why do we call it a
Law of Gravitation at all?

It will be remembered
that a space,
of any number of dimensions,
is characterized by
its expression for ds^2 (see page 187).
Thus
(56) is completely determined by
the nature of the space which,
by the Principle of Equivalence
determines the path
of a freely moving object
in the space.

*See p. 242.

217

But, even granting the
Principle of Equivalence,
that is,
granting the idea
that the nature of the space,
rather than a "force" of gravity,
determines how objects (or light)
move in that space —
in other words,
granting that the g's alone
determine the Law of Gravitation —
one may still ask:
Why is this particular expression (56)
taken to be the
Law of Gravitation?

To which the answer is that
it is the SIMPLEST expression which is
ANALOGOUS to Newton's Law of Gravitation.
Perhaps the reader is unpleasantly surprised
at this reply,
and thinks that the choice has been
made rather ARBITRARILY!
May we therefore suggest to him
to read through the rest of this book
in order to find out
the CONSEQUENCES of Einstein's choice
of the Law of Gravitation.
We predict that he will be convinced
of the WISDOM of this choice,*
and will appreciate that this is
part of Einstein's GENIUS!

*The reader who is particularly
 interested in this point
 may wish to look at d'Inverno, Chapter 10,
 in Further Reading.

He will see, for example, on page 271,
that the equations giving
the path of a planet,
derived by Newton,
are the SAME, to a first approximation,
as the Einstein equations,
so that the latter can do
ALL that the Newtonian equations do,
and FURTHERMORE,
the ADDITIONAL term in (84)
accounts for the "unusual" path
of the planet Mercury,
which the Newtonian equation (85)
did not account for at all.
But we are anticipating the story!

Let us now express Newton's Law in
a form which will show the analogy clearly.

XXVI. COMPARISON OF EINSTEIN'S LAW OF GRAVITATION WITH NEWTON'S.

Everyone knows that,
according to Newton,*
two bodies attract each other
with a force which is proportional
to the product of their masses,
and inversely proportional to the
square of the distance between them,
thus:

$$F = \frac{k m_1 m_2}{r^2}.$$

*See Holton & Brush, Chapter 11,
in Further Reading.

In this formula
we regard the two bodies,
of masses m_1 and m_2,
as each concentrated at a single point*
(its "center of gravity"),
and r is then precisely
the distance between these two points.
Now we may consider that m_1
is surrounded by a "gravitational field"
in which the gravitational force at A
(see the diagram on page 221)
is given by the above equation.
If we divide both sides by m_2
we get

$$\frac{F}{m_2} = \frac{km_1}{r^2}.$$

And, according to Newton,

$$\frac{F}{m_2} = a,$$

a is the acceleration with which
m_2 would move due to
the force F acting on it.
We may therefore write

$$a = \frac{C}{r^2} \tag{58}$$

where the constant C now includes m_1
since we are speaking of
the gravitational field around m_1.

*Thus it is a fact that
to support a body
it is not necessary to
hold it up all over,
but one needs only support it
right under its center of gravity,
as if its entire mass
were concentrated at that point.

Now, acceleration is a vector quantity,*
and it may be split up into components:†
Thus take the origin to be at m_1,
and the mass m_2 at A:

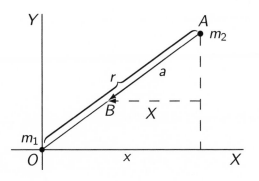

then $OA = r$;
and let AB represent the acceleration at A
(since m_2 is being pulled toward m_1)
in both magnitude and direction.
Now if X is the x-component of a,
it is obvious that

$$\frac{X}{a} = \frac{x}{r}.$$

Therefore

$$X = a \cdot \frac{x}{r}.$$

Or, better,

$$X = -a \cdot \frac{x}{r}$$

*See page 127.
†See page 129.

221

to show that the direction of X is to the left.
Substituting in this equation
the value of a from (58)
we get:

$$X = -\frac{Cx}{r^3}.$$

And, similarly,

$$Y = -\frac{Cy}{r^3},$$

and, in 3-dimensional space,
we would have also

$$Z = -\frac{Cz}{r^3}.$$

By differentiation, we get:

$$\frac{\partial X}{\partial x} = \frac{-Cr^3 + 3Cr^2x \cdot \partial r/\partial x}{r^6}.$$

But, since $r^2 = x^2 + y^2 + z^2$
(as is obvious from the diagram
on page 131, if $AB = r$),
then

$$\frac{\partial r}{\partial x} = \frac{x}{r}.$$

Substituting this in the above equation,
it becomes

$$\frac{\partial X}{\partial x} = \frac{-Cr^3 + 3Cx^2r}{r^6} = \frac{-C(r^2 - 3x^2)}{r^5}.$$

And, similarly,

$$\frac{\partial Y}{\partial y} = \frac{-C(r^2 - 3y^2)}{r^5} \quad \text{and} \quad \frac{\partial Z}{\partial z} = \frac{-C(r^2 - 3z^2)}{r^5}.$$

From these we get:

$$\frac{\partial X}{\partial x} + \frac{\partial Y}{\partial y} + \frac{\partial Z}{\partial z} = 0. \tag{59}$$

This equation may be written:

$$\frac{\partial^2 \phi}{\partial x^2} + \frac{\partial^2 \phi}{\partial y^2} + \frac{\partial^2 \phi}{\partial z^2} = 0 \qquad (60)$$

where ϕ is a function such that

$$X = \frac{\partial \phi}{\partial x}, \ Y = \frac{\partial \phi}{\partial y}, \ Z = \frac{\partial \phi}{\partial z}$$

and is called the
"gravitational potential";*
obviously (60) is merely another way
of expressing the field equation (59)
obtained from
Newton's Law of Gravitation.
This form of the law, namely (60),
is generally known as
the Laplace equation
and is more briefly denoted by

$$\nabla^2 \phi = 0$$

where the symbol ∇^2 merely denotes†
that
the second partial derivatives
with respect to x, y, and z,
respectively,
are to be taken and added together,
as shown in (60).
We see from (60), then,
that the gravitational field equation
obtained from
Newton's Law of Gravitation
is an equation containing
the second partial derivatives
of the gravitational potential.

*See d'Inverno, p. 44 & p. 124, in Further Reading.
†The symbol ∇ is read "nabla" and ∇^2 is read "nabla square". [17]

Whereas (56) is
a set of equations
which also contain
nothing higher than
the second partial derivatives
of the g's,
which,
by the Principle of Equivalence,
replace the notion of
a gravitational potential
derived from the idea of
a "force" of gravity,
by the idea of
the characteristic property of
the SPACE in question (see Ch. XII).
It is therefore reasonable
to accept (56) as the
gravitational field equations
which follow from the idea of
the Principle of Equivalence.

HOW REASONABLE it is
will be evident
when we test it by
EXPERIMENT!

It has been said (on page 215)
that each G consists of four B's.
Hence,
if the B's are all zero,
then the G's will all be zero;
but the converse
is obviously NOT true:
Namely,
even if the G's are all zero,
it does not necessarily follow
that the B's are zero.

But we know that,
to have the B's all zero
implies that
the space is Euclidean (see p. 213).

Thus,
if the condition for Euclidean space
is fulfilled,
namely,

$$B^{\alpha}_{\sigma\tau\rho} = 0,$$

then $G_{\sigma\tau} = 0$ automatically follows; thus

$$G_{\sigma\tau} = 0$$

is true in the special case of
Euclidean space.
But, more than this,
since

$$G_{\sigma\tau} = 0$$

does NOT NECESSARILY imply
that the B's are zero,
hence

$$G_{\sigma\tau} = 0$$

can be true
EVEN IF THE SPACE IS
NOT EUCLIDEAN,
namely,
in the space around a body which
creates a gravitational field.

Now all this sounds very reasonable,
but still one naturally asks:
"How can this new
Law of Gravitation
be tested EXPERIMENTALLY?"

Einstein suggested several ways
in which it might be tested.
And,
as every child now knows,
when the experiments were
actually carried out,
his predictions were all fulfilled,
and caused a great stir
not only in the scientific world,
but penetrated even into
the daily news
the world over.

But doubtless the reader
would like to know
the details of these experiments,
and just how the above-mentioned
Law of Gravitation
is applied to them.

That is what we shall show next.

XXVII. HOW CAN THE EINSTEIN LAW OF GRAVITATION BE TESTED?

We have seen that

$$G_{\sigma\tau} = 0$$

represents Einstein's new
Law of Gravitation,
and consists of 6 equations
containing partial derivatives of
the little g's.*

*See pages 215 to 217.

In order to test this law
we must obviously substitute in it
the value of the g's which
actually apply in our physical world;
in other words,
we must know first
what is the expression for ds^2
which applies to our world
(see Chapter XIII).

Now, if we use
the customary polar coordinates,
we know that
in two-dimensional EUCLIDEAN space
we have*

$$ds^2 = dr^2 + r^2 d\theta^2.$$

Similarly,
for three-dimensional
EUCLIDEAN space
we have the well-known:

$$ds^2 = dr^2 + r^2 d\theta^2 + r^2 \sin^2 \theta \cdot d\phi^2.$$

The reader can easily derive this from

$$ds^2 = dx_1^2 + dx_2^2 + dx_3^2$$

(on page 189),
by changing to polar coordinates
with the aid of the diagram
on page 230.

*See page 123.

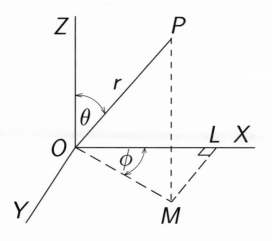

where

$$x_1 = x = OL = OM \cos\phi$$
$$= r \cos \angle POM \cos\phi = r \sin\theta \cos\phi$$
$$x_2 = y = LM = OM \sin\phi = r \sin\theta \sin\phi$$
$$x_3 = z = PM = r \cos\theta.$$

And,
for 4-dimensional space-time

230

we have

$$ds^2 = -dr^2 - r^2 d\theta^2 - r^2 \sin^2\theta \cdot d\phi^2 + c^2 dt^2$$

or

$$ds^2 = -dx_1^2 - x_1^2 dx_2^2 - x_1^2 \sin^2 x_2 \cdot dx_3^2 + dx_4^2$$

(61a)

(where $x_1 = r$, $x_2 = \theta$, $x_3 = \phi$, $x_4 = t$,
and c is taken equal to 1),
as we can readily see:

Note that the general form for
four-dimensional space
in Cartesian coordinates,
analogous to the 3-dimensional one on p. 189,
is:

$$ds^2 = dx^2 + dy^2 + dz^2 + d\tau^2.$$

But, on page 67
we showed that
in order to get
the square of an "interval" in
space-time
in this form,
with all four plus signs,
we had to take τ NOT equal to
the time, t,
BUT to take $\tau = -ict$,* where
$i = \sqrt{-1}$, and
$c =$ the velocity of light;
from which

$$d\tau^2 = -c^2 dt^2,$$

and the above expression becomes:

$$ds^2 = dx^2 + dy^2 + dz^2 - c^2 dt^2.$$

---

*As a matter of fact,
in "Special Relativity,"
we took $\tau = -it$,
but that was because
we also took $c = 1$;
otherwise, we must take $\tau = -ict$.

231

And, furthermore,
since in actual fact,
c^2dt^2 is always found to be
greater than $(dx^2 + dy^2 + dz^2)$,
therefore,
to make ds come out real instead of imaginary,
it is more reasonable to write

$$ds^2 = -dx^2 - dy^2 - dz^2 + c^2dt^2,$$

which in polar coordinates,
becomes (61a).

The reader must clearly realize that
this formula still applies to
EUCLIDEAN space-time,
which is involved in
the SPECIAL theory of Relativity*
where we considered only
observers moving with
UNIFORM velocity relative to each other.
But now,
in the GENERAL theory (page 96)
where we are considering
accelerated motion (page 102),
and therefore have a
NON-EUCLIDEAN space-time
(see Chapter XII),
what expression for ds^2
shall we use? [18]

In the first place
it is reasonable to assume that

$$ds^2 = -e^\lambda dx_1^2 - e^\mu(x_1^2 dx_2^2 + x_1^2 \sin^2 x_2\, dx_3^2) \\ + e^\nu dx_4^2. \tag{61b}$$

(where x_1, x_2, x_3, x_4 represent

*See Part I of this book.

233

the polar coordinates r, θ, ϕ, and t,
respectively,
and λ, μ, and ν are functions of x_1 only),
BECAUSE:

(A) we do not include product terms
    of the form $dx_1 \cdot dx_2$,
    or, more generally,
    of the form $dx_\sigma \, dx_\tau$ where $\sigma \neq \tau$,
    (which ARE included in (42), p. 187)
    since

from astronomical evidence
it seems that
our universe is
(a) ISOTROPIC and
(b) HOMOGENEOUS:
That is,
the distribution of matter
(the nebulae)
is the SAME
(a) IN ALL DIRECTIONS and
(b) FROM WHICHEVER POINT WE LOOK.

Now,
how does the omission of terms like

$$dx_\sigma \, dx_\tau \quad \text{where } \sigma \neq \tau$$

represent this mathematically?
Well, obviously,
a term like $dr \cdot d\theta$
(or $d\theta \cdot d\phi$ or $dr \cdot d\phi$)
would be different
for θ (or ϕ or r) positive or negative,
and consequently,
the expression for ds^2
would be different if we turn
in opposite directions —

234

which would contradict the
experimental evidence that
the universe is ISOTROPIC.
And of course the use of
the same expression for ds^2
from ANY point
reflects the idea of HOMOGENEITY.
And so we see that it is reasonable
to have in (61b)
only terms involving $d\theta^2$, $d\phi^2$, dr^2,
in which it makes no difference
whether we substitute $+d\theta$ or $-d\theta$, etc.

Similarly,
since in getting a measure for ds^2,
we are considering
a STATIC condition,
and not one which is changing
from moment to moment,
we must therefore not include
terms which will have different values
for $+dt$ and $-dt$;
in other words,
we must not include
product terms like $dr \cdot dt$, etc.
In short
we must not have any terms involving

$$dx_\sigma \cdot dx_\tau \quad \text{where } \sigma \neq \tau,$$

but only terms involving

$$dx_\sigma \cdot dx_\tau \quad \text{where } \sigma = \tau.$$

(B) The factors e^λ, e^μ, e^ν, are inserted
   in the coefficients*
   to allow for the fact

   *Cf. (61a) and (61b).

235

that our space is now
NON-EUCLIDEAN.
Hence they are so chosen as to
allow freedom to adjust them
to the actual physical world
(since they are variables),
and yet
their FORM is such that
it will be easy to manipulate them
in making the necessary adjustments —
as we shall see.*

Now,
(61b) can be somewhat simplified
by replacing

$$e^{\mu} x_1^2 \text{ by } (x_1')^2,$$

and taking x_1' as a new coordinate,
thus getting rid of e^{μ} entirely;
and we may even drop the prime,
since any change in dx_1^2 which arises
from the above substitution
can be taken care of
by taking λ correspondingly different.
Thus (61b) becomes, more simply,

$$ds^2 = -e^{\lambda} \cdot dx_1^2 - x_1^2 dx_2^2 - x_1^2 \sin^2 x_2 dx_3^2 + e^{\nu} \cdot dx_4^2. \quad (62)$$

And we now have to find
the values of the coefficients

$$e^{\lambda} \text{ and } e^{\nu}$$

in terms of x_1.†

*Further justification for (61b)
may be found in
d'Inverno, Chapter 14, pp. 184–186.

†See page 234.

We warn the reader that this is a
COLOSSAL UNDERTAKING,
but, in spite of this bad news,
we hasten to provide
the consolation that
many terms will reduce to zero,
and the whole complicated structure
will melt down to almost nothing;
we can then apply the result
to the physical data
with the greatest ease.
To any reader who "can't take it"
we suggest skipping the next chapter
and merely using the result
to follow
the experimental tests of the
Einstein Law of Gravitation
given from page 255 on.

BUT THE READER WILL MISS A LOT OF FUN!

XXVIII. SURMOUNTING THE DIFFICULTIES.

So far, then,
we have the following values:

$$g_{11} = -e^{\lambda}, \ g_{22} = -x_1^2, \ g_{33} = -x_1^2 \sin^2 x_2, \ g_{44} = e^{\nu}$$

and $g_{\sigma\tau} = 0$ when $\sigma \neq \tau$. (See (62) on p. 236.)
Furthermore,
the determinant g (see page 194)
is simply equal to
the product of the four elements in
its principal diagonal,

since all the other elements are zero:

$$\begin{vmatrix} -e^\lambda & 0 & 0 & 0 \\ 0 & -x_1^2 & 0 & 0 \\ 0 & 0 & -x_1^2 \sin^2 x_2 & 0 \\ 0 & 0 & 0 & e^\nu \end{vmatrix}$$

Hence [19]

$$g = -e^{\lambda+\nu} \cdot x_1^4 \sin^2 x_2.$$

Also, in this case,*

$$g^{\sigma\sigma} = 1/g_{\sigma\sigma}$$

and

$$g^{\sigma\tau} = 0 \text{ when } \sigma \neq \tau.$$

We shall need these relationships
in determining e^λ and e^ν in (62).

Now we shall see
how the big G's will help us to
find the little g's
and how the little g's will help us
to reduce the number of big G's to
ONLY THREE!

First let us show that
the set of quantities

$$G_{\sigma\tau}$$

is SYMMETRIC,†
and therefore

---

*See the definition of $g^{\mu\nu}$ on page 196.
Note that in the equation $g^{\sigma\sigma} = 1/g_{\sigma\sigma}$
there is NO summation to be done!
This equation means: $g^{11} = 1/g_{11}$, $g^{22} = 1/g_{22}$,
and so on.
†See page 193.

$G_{\sigma\tau} = 0$ reduces to TEN equations*
instead of sixteen,
as σ and τ each take on
their values $1, 2, 3, 4$.
To show this,
we must remember that
$G_{\sigma\tau}$ really represents (57) on p. 216;
and let us examine $\{\sigma\alpha, \alpha\}$
which occurs in (57):
By definition (page 196),

$$\{\sigma\alpha, \alpha\} = \tfrac{1}{2} g^{\alpha\epsilon} \left(\frac{\partial g_{\sigma\epsilon}}{\partial x_\alpha} + \frac{\partial g_{\alpha\epsilon}}{\partial x_\sigma} - \frac{\partial g_{\sigma\alpha}}{\partial x_\epsilon} \right).$$

But,
remembering that
the presence of α and ϵ TWICE
in EACH term
(after multiplying out)
implies that we must SUM on α and ϵ,
the reader will easily see that
many of the terms will cancel out
and that we shall get

$$\{\sigma\alpha, \alpha\} = \tfrac{1}{2} g^{\alpha\epsilon} \frac{\partial g_{\alpha\epsilon}}{\partial x_\sigma}.$$

Furthermore,
by the definition of $g^{\mu\nu}$ on page 196,
the reader may also verify the fact that

$$\tfrac{1}{2} g^{\alpha\epsilon} \frac{\partial g_{\alpha\epsilon}}{\partial x_\sigma} = \frac{1}{2g} \cdot \frac{\partial g}{\partial x_\sigma}$$

where g is the determinant of p. 239.
And, from elementary calculus,

*See page 193.

240

$$\frac{1}{2g} \cdot \frac{\partial g}{\partial x_\sigma} = \frac{\partial}{\partial x_\sigma} \log \sqrt{-g}.*$$

Hence,

$$\{\sigma\alpha, \alpha\} = \frac{\partial}{\partial x_\sigma} \log \sqrt{-g}.$$

Similarly,

$$\{\epsilon\alpha, \alpha\} = \frac{\partial}{\partial x_\epsilon} \log \sqrt{-g}.$$

Substituting these values in (57),
we get:

$$G_{\sigma\tau} \equiv \{\sigma\alpha, \epsilon\} \{\epsilon\tau, \alpha\} + \frac{\partial^2}{\partial x_\sigma \cdot \partial x_\tau} \log \sqrt{-g}$$
$$- \frac{\partial}{\partial x_\alpha} \{\sigma\tau, \alpha\} - \{\sigma\tau, \epsilon\} \frac{\partial}{\partial x_\epsilon} \log \sqrt{-g} \qquad (63)$$
$$= 0.$$

We can now easily see
that (63) represents
10 equations and not sixteen,
for the following reasons:
In the first place,

$$\{\epsilon\tau, \alpha\} = \{\tau\epsilon, \alpha\} \qquad \text{(see pp. 204, 205)}.$$

Hence,
by interchanging σ and τ,
the first term of (63) remains unchanged,
its two factors merely change places

*Note that
we might also have obtained $\sqrt{+g}$,
but since g is always negative
(we shall show on p. 252 that $\lambda = -\nu$,
and therefore g on p. 239 becomes $-x_1^4 \cdot \sin^2 x_2$)
it is more reasonable to select $\sqrt{-g}$, which
will make the Christoffel symbols,
and hence also the terms in
the new Law of Gravitation,
REAL rather than imaginary.

(since ϵ and α are mere dummies,
as explained on page 204).
And,
the second, third, and fourth terms of (63)
are also unchanged by
the interchange of σ and τ.
In other words,

$$G_{\sigma\tau} = G_{\tau\sigma}.$$

Thus, if we arrange
the 16 quantities in $G_{\sigma\tau}$
in a square array:

$$\left\| \begin{array}{cccc} G_{11} & G_{12} & G_{13} & G_{14} \\ G_{21} & G_{22} & G_{23} & G_{24} \\ G_{31} & G_{32} & G_{33} & G_{34} \\ G_{41} & G_{42} & G_{43} & G_{44} \end{array} \right\|$$

We have just shown that
this is a SYMMETRIC matrix.*
Hence (63) reduces to 10 equations
instead of 16,
as we said before.

We shall not burden the reader
with the details of
how (63) is further reduced
to only SIX equations.†
But perhaps the reader is thinking
that "only six" equations
are still no great consolation,
particularly if she realizes

*See page 239.

†The interested reader
 might take a look at
 Eddington, p. 115, or
 d'Inverno, p. 89. [20]

how long each of these equations is!
But does she realize this?
She would do well to take
particular values of σ and τ,
say $\sigma = 1$, and $\tau = 1$,
in order to see just what
ONE of the equations in (63)
is really like!
(Don't forget to sum on the dummies!)

Is the reader wondering
just what we are trying to do to her?
Is this a subtle mental torture
by which we
alternatively frighten and console her?
The fact is that
we do want to frighten her sufficiently
to make her realize
the colossal amount of computation
that is involved here,
and yet to keep up her courage too
by the knowledge that
it does eventually boil down
to a really simple form.
She might not appreciate
the final simple form
if she did not know
the labor that produced it.
With this apology,
we shall now proceed to indicate
how the further simplification
takes place.

In each Christoffel symbol in (63),
we must substitute specific values
for the Greek letters.
It is obvious then

that there will be four possible types:

(a) those in which the values of
all three Greek letters are alike:
Thus: $\{\sigma\sigma,\sigma\}$

(b) those of the form $\{\sigma\sigma,\tau\}$
(c) those of the form $\{\sigma\tau,\tau\}$
and

(d) those of the form $\{\sigma\tau,\rho\}$.

Note that it is unnecessary to consider
the form $\{\tau\sigma,\tau\}$
since this is the same as $\{\sigma\tau,\tau\}$ (see p. 204).

Now, by definition (page 196),

$$\{\sigma\sigma,\sigma\} = \tfrac{1}{2} g^{\sigma\alpha}\left(\frac{\partial g_{\sigma\alpha}}{\partial x_\sigma} + \frac{\partial g_{\sigma\alpha}}{\partial x_\sigma} - \frac{\partial g_{\sigma\sigma}}{\partial x_\alpha}\right)$$

and, as usual,
we must sum on α.
But since the only g's which are not zero
are those in which
the indices are alike (see p. 237)
and, in that case,

$$g^{\sigma\sigma} = 1/g_{\sigma\sigma} \qquad \text{(p. 239).}$$

Hence

$$\{\sigma\sigma,\sigma\} = \frac{1}{2g_{\sigma\sigma}}\left(\frac{\partial g_{\sigma\sigma}}{\partial x_\sigma} + \frac{\partial g_{\sigma\sigma}}{\partial x_\sigma} - \frac{\partial g_{\sigma\sigma}}{\partial x_\sigma}\right)$$

and therefore
$$\{\sigma\sigma,\sigma\} = \frac{1}{2g_{\sigma\sigma}} \cdot \frac{\partial g_{\sigma\sigma}}{\partial x_\sigma}$$

which, by elementary calculus, gives

(a) $\qquad \{\sigma\sigma,\sigma\} = \tfrac{1}{2}\dfrac{\partial}{\partial x_\sigma}\log g_{\sigma\sigma}.$

Similarly,

$$\{\sigma\sigma, \tau\} = \tfrac{1}{2} g^{\tau\alpha} \left(\frac{\partial g_{\sigma\alpha}}{\partial x_\sigma} + \frac{\partial g_{\sigma\alpha}}{\partial x_\sigma} - \frac{\partial g_{\sigma\sigma}}{\partial x_\alpha} \right).$$

Here the only values of α that
will keep the outside factor $g^{\tau\alpha}$ from being zero
are those for which $\alpha = \tau$,
and since $\tau \neq \sigma$
(for otherwise we should have case (a))
we get

$$\{\sigma\sigma, \tau\} = -\tfrac{1}{2} g^{\tau\tau} \cdot \frac{\partial g_{\sigma\sigma}}{\partial x_\tau}$$

or

(b) $$\{\sigma\sigma, \tau\} = -\frac{1}{2g_{\tau\tau}} \cdot \frac{\partial g_{\sigma\sigma}}{\partial x_\tau}.$$

Likewise

(c) $$\{\sigma\tau, \tau\} = \tfrac{1}{2} \frac{\partial}{\partial x_\sigma} \log g_{\tau\tau}$$

and

(d) $$\{\sigma\tau, \rho\} = 0.$$

Let us now evaluate these various forms
for specific values:
Thus, take, in case (a), $\sigma = 1$:

Then $$\{11, 1\} = \tfrac{1}{2} \cdot \frac{\partial}{\partial x_1} \log g_{11}$$

But $g_{11} = -e^\lambda$ (see p. 239).

Hence $\{11, 1\} = \tfrac{1}{2} \cdot \dfrac{\partial}{\partial x_1} \log(-e^\lambda)$

which, by elementary calculus, gives

$$\{11, 1\} = \tfrac{1}{2} \left(\frac{-e^\lambda}{-e^\lambda} \right) \frac{\partial \lambda}{\partial x_1} = \tfrac{1}{2} \cdot \frac{\partial \lambda}{\partial x_1} = \tfrac{1}{2} \lambda',$$

where λ' represents $\dfrac{\partial \lambda}{\partial x_1}$ or $\dfrac{\partial \lambda}{\partial r}$,

since $x_1 = r$ (see page 233).

Similarly,

$$\{22, 2\} = \tfrac{1}{2} \cdot \frac{\partial}{\partial x_2} \log g_{22} = \tfrac{1}{2} \frac{\partial}{\partial x_2} \log(-x_1^2).$$

But, since
in taking a PARTIAL derivative
with respect to one variable,
all the other variables are held constant,
hence

$$\frac{\partial}{\partial x_2} \log(-x_1^2) = 0,$$

and therefore

$$\{22, 2\} = 0.$$

And, likewise,

$$\{33, 3\} = \{44, 4\} = 0.$$

Now, for case (b),
take first $\sigma = 1, \tau = 2$;
then

$$\{11, 2\} = -\frac{1}{2g_{22}} \cdot \frac{\partial}{\partial x_2} g_{11} = -\frac{1}{2g_{22}} \cdot \frac{\partial}{\partial x_2} \left(-e^{\lambda}\right).$$

But, since λ is a function of x_1 only,*
and is therefore held constant
while the partial derivative
with respect to x_2 is taken,
hence $\{11, 2\} = 0$,
and so on.

Let us see how many specific values
we shall have in all.
Obviously (a) has 4 specific cases,

*See page 234.

namely, $\sigma = 1, 2, 3, 4,$
which have already been evaluated above.

(b) will have 12 specific cases,
since for each value of $\sigma = 1, 2, 3, 4,$
τ can have 3 of its possible 4 values
(for here $\sigma \neq \tau$);
(c) will also have 12 cases,
and
(d) will have $4 \times 3 \times 2 = 24$ cases,
but since $\{\sigma\tau, \rho\} = \{\tau\sigma, \rho\}$ (see p. 204),
this reduces to 12.

Hence in all
there are 40 cases.

The reader should verify the fact that
31 of the 40 reduce to zero,
the 9 remaining ones being*

$$\left.\begin{aligned}
\{11, 1\} &= \tfrac{1}{2}\lambda' \\[6pt]
\{12, 2\} &= \frac{1}{r} \\[6pt]
\{13, 3\} &= \frac{1}{r} \\[6pt]
\{14, 4\} &= \tfrac{1}{2}\nu' \\[6pt]
\{22, 1\} &= -re^{-\lambda} \\[6pt]
\{23, 3\} &= \cot\theta \\[6pt]
\{33, 1\} &= -r\sin^2\theta \cdot e^{-\lambda} \\[6pt]
\{33, 2\} &= -\sin\theta \cdot \cos\theta \\[6pt]
\{44, 1\} &= \tfrac{1}{2}e^{\nu-\lambda} \cdot \nu'
\end{aligned}\right\} \qquad (64)$$

*Remember that $x_2 = \theta$: see page 233.

Note that $\nu' = \dfrac{\partial \nu}{\partial x_1} = \dfrac{\partial \nu}{\partial r}$.

Now, in (63),
when we give to the various Greek letters
their possible values,
we find that,
since so many of the Christoffel symbols
are equal to zero,
a great many (over 200) terms drop out!
And there remain now
only FIVE equations,
each with a much smaller number of terms.
These are written out in full below,
and,
lest the reader think that
this is the promised,
final simplified result,
we hasten to add that
the BEST is yet to come!

Just how G_{11} is obtained,
showing the reader how
to SUM on α and ϵ
and which terms drop out
(because they contain zero factors)
will be found in V,
on page 317.
And,
similarly for the other G's.
Here we give
the equations which result
after the zero terms have been
eliminated.

$$G_{11} = \{11,1\}\{11,1\} + \{12,2\}\{21,2\} +$$
$$\{13,3\}\{31,3\} + \{14,4\}\{41,4\}$$
$$- \frac{\partial}{\partial x_1}\{11,1\} + \frac{\partial^2}{\partial x_1^2} \log\sqrt{-g}$$
$$- \{11,1\}\frac{\partial}{\partial x_1} \log\sqrt{-g}$$
$$= 0.$$

Similarly,

$$G_{22} = 2\{22,1\}\{12,2\} + \{23,3\}\{23,3\}$$
$$- \frac{\partial}{\partial x_1}\{22,1\} + \frac{\partial^2}{\partial x_2^2} \log\sqrt{-g}$$
$$- \{22,1\}\frac{\partial}{\partial x_1} \log\sqrt{-g}$$
$$= 0.$$

$$G_{33} = 2\{33,1\}\{13,3\} + 2\{33,2\}\{23,3\}$$
$$- \frac{\partial}{\partial x_1}\{33,1\} - \frac{\partial}{\partial x_2}\{33,2\}$$
$$- \{33,1\}\frac{\partial}{\partial x_1} \log\sqrt{-g}$$
$$- \{33,2\}\frac{\partial}{\partial x_2} \log\sqrt{-g}$$
$$= 0.$$

$$G_{44} = 2\{44,1\}\{14,4\} - \frac{\partial}{\partial x_1}\{44,1\}$$
$$- \{44,1\}\frac{\partial}{\partial x_1} \log\sqrt{-g}$$
$$= 0.$$

$$G_{12} = \{13,3\}\{23,3\} - \{12,2\}\frac{\partial}{\partial x_2} \log\sqrt{-g}$$
$$= 0.$$

If we now substitute
in these equations
the values given in (64),
we get*

$$G_{11} = \tfrac{1}{4}\lambda'^2 + \frac{1}{r^2} + \frac{1}{r^2} + \tfrac{1}{4}\nu'^2 - \tfrac{1}{2}\lambda'' + \left(\tfrac{1}{2}\lambda''\right.$$

$$+ \left. \tfrac{1}{2}\nu'' - \frac{2}{r^2}\right) - \tfrac{1}{2}\lambda'\left(\tfrac{1}{2}\lambda' + \tfrac{1}{2}\nu' + \frac{2}{r}\right)$$

$$= \tfrac{1}{4}\nu'^2 + \tfrac{1}{2}\nu'' - \tfrac{1}{4}\lambda'\nu' - \frac{\lambda'}{r}$$

$$= 0.$$

Similarly

$$G_{22} = e^{-\lambda}\left[1 + \tfrac{1}{2}r\left(\nu' - \lambda'\right)\right] - 1$$

$$= 0.$$

$$G_{33} = \sin^2\theta \cdot e^{-\lambda}\left[1 + \tfrac{1}{2}r\left(\nu' - \lambda'\right)\right] - \sin^2\theta$$

$$= 0.$$

$$G_{44} = e^{\nu-\lambda}\left(-\tfrac{1}{2}\nu'' + \tfrac{1}{4}\lambda'\nu' - \tfrac{1}{4}\nu'^2 - \frac{\nu'}{r}\right)$$

$$= 0.$$

*Here $\quad \lambda'' = \dfrac{\partial^2\lambda}{\partial r^2}$

and $\quad \nu'' = \dfrac{\partial^2\nu}{\partial r^2}$

and G_{12} becomes:

$$\frac{1}{r}\cot\theta - \frac{1}{r}\cot\theta = 0$$

which is identically zero
and therefore drops out,
thus reducing the number of equations
to FOUR.

Note also that
G_{33} includes G_{22},
so that these two equations
are not independent —
hence now the equations are
THREE.

And now, dividing G_{44} by $e^{\nu-\lambda}$
and adding the result to G_{11},
we get

$$\lambda' = -\nu' \qquad\qquad (65)$$

or

$$\frac{\partial\lambda}{\partial r} = -\frac{\partial\nu}{\partial r}.$$

Therefore, by integration,

$$\lambda = -\nu + C \qquad\qquad (66)$$

where C is a constant of integration.

But, since
at an infinite distance from matter,
our universe would be Euclidean,*

*See page 226.

and then, for Cartesian coordinates,
we would have*:

$$ds^2 = dx_1^2 + dx_2^2 + dx_3^2 + dx_4^2,$$

that is,
the coefficient of dx_1^2 and dx_4^2
must be 1 under these conditions;
hence,
if (61b) is to hold also for
this special case,
as of course it must do,
we should then have $\lambda = 0$, $\nu = 0$.

In other words,
since,
when $\nu = 0$, λ also equals 0,
then, from (66), C, too, must be zero.
Hence

$$\lambda = -\nu. \tag{67}$$

Using (65) and (67),
G_{22} on page 250 becomes

$$e^\nu \left(1 + r\nu'\right) = 1. \tag{68}$$

If we put $\gamma = e^\nu$,
and differentiate with respect to r,
we get

$$\frac{\partial \gamma}{\partial r} = e^\nu \cdot \frac{\partial \nu}{\partial r}$$

or

$$\gamma' = e^\nu \cdot \nu'.$$

Hence (68) becomes

$$\gamma + r\gamma' = 1. \tag{69}$$

*See pages 189 and 231.

This equation may now be
easily integrated,*
obtaining

$$\gamma = 1 - \frac{2m}{r} \tag{70}$$

where $2m$ is a constant of integration.
The constant m will later be shown
to have an important physical meaning.

Thus we have succeeded in finding

$$e^\lambda \text{ and } e^\nu$$

---

*From the elementary theory of
differential equations,
we write (69):

$$\gamma + r\frac{d\gamma}{dr} = 1$$

or

$$r\frac{d\gamma}{dr} = 1 - \gamma$$

or

$$-\frac{d\gamma}{1-\gamma} = -\frac{dr}{r}.$$

Having separated the variables,
we can now integrate both sides
thus:

$$\log(1 - \gamma) = -\log r + \text{constant};$$

or

$$\log r(1 - \gamma) = \text{constant},$$

and therefore

$$r(1 - \gamma) = \text{constant}.$$

We may for convenience
let the constant equal $2m$,
and write

$$r(1 - \gamma) = 2m,$$

from which we get

$$\gamma = 1 - \frac{2m}{r}.$$

in terms of x_1:

$$e^\nu = 1/e^\lambda = \gamma = 1 - \frac{2m}{r} = 1 - \frac{2m}{x_1}$$

and (62) becomes:

$$ds^2 = -\gamma^{-1}dr^2 - r^2 d\theta^2 - r^2 \sin^2 \theta \cdot d\phi^2 + \gamma dt^2, \qquad (71)$$

where, as before (p. 233),

$$r = x_1, \quad \theta = x_2, \quad \phi = x_3, \quad t = x_4.$$

And hence
the new Law of Gravitation,
consisting now of only
the THREE remaining equations:

$$G_{11} = 0, \ G_{33} = 0, \text{ and } G_{44} = 0,$$

are now fully determined by the
the little g's of (71).

We can now proceed
to test this result
to see whether it really applies
to the physical world we live in.

XXIX. "THE PROOF OF THE PUDDING."

The first test is naturally
to see what
the new Law of Gravitation
has to say about
the path of a planet.
It was assumed by Newton that
a body "naturally" moves
along a straight line

255

if it is not pulled out of its course
by some force acting on it:
As, for example,
a body moving on
a frictionless Euclidean plane.
Similarly,
according to Einstein
if it moves "freely" on
the surface of a SPHERE
it would go along the
"nearest thing to a straight line,"
that is,
along the GEODESIC for this surface,
namely,
along a great circle.
And, for other surfaces,
or spaces of higher dimensions,
it would move along
the corresponding geodesic for
the particular surface or space.

Now our problem is
to find out
what is the geodesic in
our non-Euclidean physical world,
since a planet must move
along such a geodesic.

In order to find
the equation of a geodesic
it is necessary to know
something about the
"Calculus of Variations," [21]
so that we cannot go into details here.
But we shall give the reader
a rough idea of the plan,
together with references where

he may look up this matter further.*
Suppose, for example, that
we have given
two points, A and B, on a
Euclidean plane;
it is obviously possible to
join them by various paths,
thus:

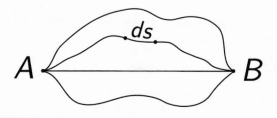

Now,
which of all possible paths
is the geodesic here?
Of course the reader knows the answer:
It is the straight line path.
But how do we set up
the problem mathematically
so that we may solve
similar problems in other cases?

*See Eddington, pp. 59–60;
d'Inverno, pp. 82–83, pp. 99–101;
Lawden [12], §43, "Geodesics," pp. 114–117.

Well,
we know from ordinary calculus
that
if a short arc on
any of these paths
is represented by ds,
then

$$\int_A^B ds$$

represents the total length of
that entire path.
And this of course applies to
any one of the paths from A to B.
How do we now select from among these
the geodesic?

This problem is similar to
one with which the reader is
undoubtedly familiar,
namely,

if $y = f(x)$,
find the values of y for which
y is an "extremum" or a "stationary."

Such values of y are shown in
the diagram on the next page
at a, b, and c:

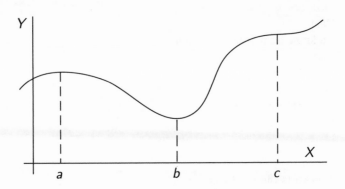

and, for all these, we must have

$$dy/dx = 0$$

or

$$dy = f'(x_o) \cdot dx = 0 \qquad (72)$$

where x_o is a, b, or c.

Similarly,
to go back to our problem on pp. 258 and 259,
the geodesic we are looking for
would make

$$\int_A^B ds$$

a stationary.
This is expressed in
the calculus of variations by

$$\delta \int_A^B ds = 0 \qquad (73)$$

analogously to (72).

To find the equation of the geodesic,
satisfying (73),
is not as simple as finding

a maximum or minimum in
ordinary calculus,
and we shall give here
only the result:*

$$\frac{d^2 x_\sigma}{ds^2} + \{\alpha\beta, \sigma\} \frac{dx_\alpha}{ds} \cdot \frac{dx_\beta}{ds} = 0 \qquad (74)$$

Let us consider (74):
In the first place,
for ordinary three-dimensional
Euclidean space
σ would have
three possible values: 1, 2, 3,
since we have here
three coordinates x_1, x_2, x_3;
furthermore,
by choosing Cartesian coordinates,
we would have (see page 189):

$$g_{11} = g_{22} = g_{33} = 1$$

and

$$g_{\mu\nu} = 0 \quad \text{for } \mu \neq \nu$$

and therefore

$$\{\alpha\beta, \sigma\}$$

which involves derivatives of the g's†
would be equal to zero,
so that (74) would become

$$\frac{d^2 x_\sigma}{ds^2} = 0 \qquad (\sigma = 1, 2, 3). \qquad (75)$$

*For details, see the references in
  the footnote on p. 258.

†See (46) on p. 196.

Now, if in (75)
we replace ds by dt,
it becomes*

$$\frac{d^2 x_\sigma}{dt^2} = 0 \qquad (\sigma = 1, 2, 3) \qquad (76)$$

which is a short way of writing
the three equations:

$$\frac{d^2 x_1}{dt^2} = 0, \qquad \frac{d^2 x_2}{dt^2} = 0, \qquad \frac{d^2 x_3}{dt^2} = 0. \qquad (77)$$

But what is
the PHYSICAL MEANING of (77)?

---

*If we consider an observer who
has chosen her coordinates
in such a way that

$$dx_1 = dx_2 = dx_3 = 0,$$

in other words,
an observer who is traveling with
a moving object,
and for whom the object is therefore
standing still with reference to
her ordinary space-coordinates,
so that only time is changing for her,
then, for her (61a) becomes

$$ds^2 = dx_4^2$$

or
$$ds^2 = dt^2$$
or
$$ds = dt.$$

That is to say,
ds becomes of the nature of "time";
for this reason,
ds is often called
"the proper time"
since it is a "time"
for the moving object itself.

Why, everyone knows that,
for uniform motion,

$$v = s/t,$$

where v is the velocity with which
a body moves when
it goes a distance of s feet in
t seconds.
If the motion is NOT uniform,
we can, by means of
elementary differential calculus,
express the velocity AT AN INSTANT,
by

$$v = ds/dt.$$

Or, if x, y, and z are
the projections of s on the
X, Y, and Z axes, respectively,
and v_x, v_y, and v_z are
the projections of v
on the three axes,
then

$$v_x = \frac{dx}{dt}, \quad v_y = \frac{dy}{dt}, \quad v_z = \frac{dz}{dt}.$$

Or,
in the abridged notation,

$$v_\sigma = dx_\sigma/dt \qquad (\sigma = 1, 2, 3)$$

where we use x_1, x_2, x_3
instead of x, y, z,
and v_1, v_2, v_3 instead of v_x, v_y, v_z.

Furthermore,
since acceleration is
the change in velocity per unit of time,

we have

$$a = \frac{dv}{dt} \quad \text{or} \quad a = \frac{d^2s}{dt^2}$$

or

$$a_\sigma = \frac{d^2 x_\sigma}{dt^2} \qquad (\sigma = 1, 2, 3). \tag{78}$$

Thus (77) states that
the components of the acceleration
must be zero,
and hence the acceleration itself
must be zero, thus:

$$a = \frac{d^2s}{dt^2} = 0$$

or

$$a = \frac{dv}{dt} = 0.$$

From this we get, by integrating

$$v = v_o,$$

a constant equal to the initial velocity, or

$$\frac{ds}{dt} = v_o,$$

and therefore
by integrating again,

$$s = v_o t + s_o,$$

where s_o is a constant equal to the initial position;
which is the equation of
A STRAIGHT LINE,
(with slope v_o and intercept s_o).

In other words,
when the equations for a geodesic,
namely, (74),
are applied to the special case of
THREE-DIMENSIONAL EUCLIDEAN SPACE,
they lead to the fact that

264

in this special case
THE GEODESIC IS A STRAIGHT LINE!

We hope the reader is DELIGHTED
and NOT DISAPPOINTED
to get a result which is
so familiar to him;
and we hope it gives him
a friendly feeling of confidence
in (74)!
And of course he must realize
that (74) will work also
for any non-Euclidean space,
since it contains
the little g's
which characterize the space;*
and for any dimensionality,
since σ may be given
any number of values.

In particular,
in our four-dimensional
non-Euclidean world,
(74) represents
the path of an object moving
in the presence of matter
(which merely makes the space
non-Euclidean),
with no external force acting upon the object;
and hence (74) is
THE PATH OF A PLANET
which we are looking for!

*See p. 190.

XXX. MORE ABOUT THE PATH OF
A PLANET.

Of course (74) is only
a GENERAL expression,
and does not yet apply to
our particular physical world,
since the Christoffel symbol

$$\{\alpha\beta, \sigma\}$$

involves the g's,
and is therefore not specific until
we substitute the value of the g's
which apply in a specific case
in the physical world.

Now in (64)
we have the values of $\{\alpha\beta, \sigma\}$
in terms of λ, ν, r, and θ.
And, by (67), $\lambda = -\nu$,
hence we know $\{\alpha\beta, \sigma\}$ in terms of

$$\nu, r, \text{ and } \theta.$$

Further, since $e^\nu = \gamma$ (see page 252)
and γ is known in terms of r from (70),
we therefore have $\{\alpha\beta, \sigma\}$ in terms of

$$r \text{ and } \theta.$$

The reader must bear in mind
that
whereas (76), in Newtonian physics,
represents only three equations,
on the other hand,
(74) in Einsteinian physics
is an abridged notation for
FOUR equations,

as σ takes on
its FOUR possible values: 1, 2, 3, 4.
Taking first the value $\sigma = 2$,
and, remembering that $x_2 = \theta$ (see page 233),
we have,
for one of the equations of (74),
the following:

$$\frac{d^2\theta}{ds^2} + \{\alpha\beta, \sigma\} \frac{dx_\alpha}{ds} \cdot \frac{dx_\beta}{ds} = 0. \qquad (79)$$

And now,
since α and β each occur
TWICE
in the second term,
we must sum on these as usual,
so that we must consider terms
containing, respectively,

$$\{11, 2\}, \ \{12, 2\}, \ \{13, 2\}, \ \{14, 2\},$$
$$\{21, 2\}, \ \{22, 2\}, \ \{23, 2\}, \ \{24, 2\}, \text{ etc.}$$

in which σ always equals 2,
and α and β each runs its course
from 1 to 4.
But, from (64), we see that
most of these are zero,
the only ones remaining being

$$\{12, 2\} = \frac{1}{r}$$

and

$$\{33, 2\} = -\sin\theta \cdot \cos\theta.$$

Also, by page 204,

$$\{21, 2\} = \{12, 2\}.$$

267

Thus (79) becomes

$$\frac{d^2\theta}{ds^2} + \frac{2}{r} \cdot \frac{dr}{ds} \cdot \frac{d\theta}{ds} - \sin\theta \cdot \cos\theta \left(\frac{d\phi}{ds}\right)^2 = 0. \qquad (80)$$

If we now choose our coordinates
in such a way
that
an object begins moving in the plane

$$\theta = \pi/2,$$

then

$$\frac{d\theta}{ds} = 0 \ \text{ and } \cos\theta = 0$$

and hence

$$\frac{d^2\theta}{ds^2} = 0.$$

If we now substitute all these values in (80),
we see that this equation is satisfied,
and hence $\theta = \pi/2$ is a solution of the equation,
thus showing that
the path of the planet
is in a plane.

Thus from (80)
we have found out that
a planet,
according to Einstein,
must move in a plane,
just as in Newtonian physics.

Let us now examine (74) further,
and see what
the 3 remaining equations in it
tell us about planetary motion:

For $\sigma = 1$,
(79) becomes

$$\frac{d^2 x_1}{ds^2} + \{11, 1\} \left(\frac{dx_1}{ds}\right)^2 + \{22, 1\} \left(\frac{dx_2}{ds}\right)^2$$
$$+ \{33, 1\} \left(\frac{dx_3}{ds}\right)^2 + \{44, 1\} \left(\frac{dx_4}{ds}\right)^2 = 0.$$

Or

$$\frac{d^2 r}{ds^2} + \tfrac{1}{2}\lambda' \left(\frac{dr}{ds}\right)^2 - re^{-\lambda} \left(\frac{d\theta}{ds}\right)^2$$
$$- r \cdot \sin^2 \theta \cdot e^{-\lambda} \left(\frac{d\phi}{ds}\right)^2 + \tfrac{1}{2}e^{\nu-\lambda} \cdot \nu' \left(\frac{dt}{ds}\right)^2 = 0.$$

But since we have chosen $\theta = \pi/2$,
then
$$\frac{d\theta}{ds} = 0 \quad \text{and} \quad \sin \theta = 1,$$

hence this equation becomes

$$\frac{d^2 r}{ds^2} + \tfrac{1}{2}\lambda' \left(\frac{dr}{ds}\right)^2 - re^{-\lambda} \left(\frac{d\phi}{ds}\right)^2$$
$$+ \tfrac{1}{2}e^{\nu-\lambda} \cdot \nu' \left(\frac{dt}{ds}\right)^2 = 0. \tag{81}$$

And similarly,
for $\sigma = 3$,
(79) gives

$$\frac{d^2 \phi}{ds^2} + \frac{2}{r} \cdot \frac{dr}{ds} \cdot \frac{d\phi}{ds} = 0, \tag{82}$$

and for $\sigma = 4$,
we get

$$\frac{d^2 t}{ds^2} + \nu' \cdot \frac{dr}{ds} \cdot \frac{dt}{ds} = 0. \tag{83}$$

269

And now,
from (81), (82), (83), and (71),
together with (70),
we get*

$$\left.\begin{array}{c} \dfrac{d^2u}{d\phi^2}+u = \dfrac{m}{h^2} + 3mu^2 \\[2ex] r^2\,\dfrac{d\phi}{ds} = h \end{array}\right\} \tag{84}$$

where m and h are
constants of integration,
and $u = 1/r$.

Thus (84) represents
the path of an object moving freely,
that is,
not constrained by any external force,
and is therefore,
in a sense,
analogous to a straight line in
Newtonian physics.
But it must be remembered
that in Einsteinian physics,
owing to the
Principle of Equivalence (Chapter XI),
an object is
NOT constrained by any external force
even when it is moving in
the presence of matter,
as, for example,
a planet moving
around the sun.
And hence (84)
would represent the path of a planet.

*For details see Eddington, p. 86;
 or d'Inverno, pp. 195–196. [22]

From this point of view
we are not interested in
comparing (84) with
the straight line motion in
Newtonian physics,
as mentioned on page 270,
but rather with
the equations representing
the path of a planet
in Newtonian physics,
in which, of course,
the planet is supposed to move
under the
GRAVITATIONAL FORCE
of the sun.
It has been shown
in Newtonian physics
that a body
moving under a "central force,"
(like a planet moving
under the influence of the sun)
moves in an ellipse,
with the central force (the sun)
located at one of the foci. [23]

And the equations of this path are:

$$\left.\begin{array}{c} \dfrac{d^2 u}{d\phi^2} + u = \dfrac{m}{h^2} \\[2mm] r^2\, \dfrac{d\phi}{dt} = h \end{array}\right\} \qquad (85)$$

where r is the distance
from the sun to the planet,
m is the mass of the sun,
and ϕ is the angle swept out by the planet
in time t.

We notice at once
the remarkable resemblance between
(84) and (85).
They are indeed
IDENTICAL[#] EXCEPT for
the presence of the term $3mu^2$,
and of course the use of
ds instead of dt in (84).*

Thus we see that
the Newtonian equations (85)
are really
a first approximation to
the Einstein equations (84);
that is why
they worked so satisfactorily
for so long.

Let us now see
how the situation is affected by
the additional term $3mu^2$.

XXXI. THE PERIHELION OF MERCURY.

Owing to the presence of the term

$$3mu^2$$

(84) is no longer an ellipse
but a kind of spiral
in which the path
is NOT retraced
each time the planet

[#]So the integration constant m
  introduced in (70) (p. 253)
  IS an important physical quantity,
  a MASS, here the mass of the sun.
*See p. 262.

makes a complete revolution,
but is shifted as shown[#]
in the following diagram:

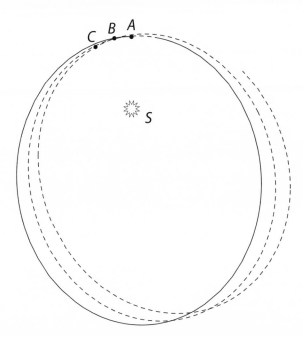

in which
the "perihelion," that is,
the point in the path
nearest the sun, S, at the focus,
is at A the first time around,
at B the next time,
at C the next,
and so on.

[#] The size of the shift is greatly exaggerated
to make the tiny effect easy to see.

In other words,
a planet does not go
round and round
in the same path,
but there is a slight shift
in the entire path,
each time around.
And the shift of the perihelion
can be calculated
by means of (84).*
This shift can also be
MEASURED experimentally,
and therefore can serve
as a method of
TESTING
the Einstein theory
in actual fact.

Now it is obvious that
when a planetary orbit
is very nearly CIRCULAR
this shift in the perihelion
is not observable,
and this is unfortunately
the situation with
most of the planets.
There is one, however,
in which this shift
IS measurable,
namely,
the planet MERCURY.

Lest the reader think
that the astronomers

*See Eddington, pp. 88–90; or
d'Inverno, pp. 195–198. [24]

274

can make only
crude measurements,
let us say in their defense,
that the discrepancy
even in the case of Mercury
is an arc of
ONLY ABOUT
43 SECONDS PER CENTURY!

Let us make clear what we mean by
"the discrepancy":
when we say that
the Newtonian theory
requires the path of a planet to be
an ellipse,
it must be understood that
this would hold only if
there were a SINGLE planet;
the presence of other planets
causes so-called "perturbations,"
so that
even according to Newton
there would be
some shift in the
perihelion.
But the amount of shift
due to this cause
has long been known to be
531 seconds of arc per century,
whereas observation shows
that the actual shift is
574 seconds,
thus leaving a shift of
43 seconds per century
UNACCOUNTED FOR
in the Newtonian theory.

Think of the DELICACY

of the measurements
and the patient persistence
over a long period of years
by generations of astronomers
that is represented
by the above figure!
And this figure was known
to astronomers
long before Einstein.
It worried them deeply
since they could not account
for the presence of this shift.

And then
the Einstein theory,
which originated in the attempt
to explain
the Michelson-Morley experiment,*
and NOT AT ALL with the intention
of explaining the shift
in the perihelion of Mercury,
QUITE INCIDENTALLY EXPLAINED
THIS DIFFICULTY ALSO,
for the presence of the term $3mu^2$ in (84)
leads to the additional shift of perihelion
of 42.9"! †

XXXII. DEFLECTION OF A RAY OF LIGHT.

We saw in the previous chapter
that the experimental evidence

*See Part I: The Special Theory of Relativity.

†See Eddington, pp. 88–90; or
  d'Inverno, pp. 195–198. [24]

in connection with
the shift of the perihelion of Mercury
was already at hand
when Einstein's theory was proposed,
and immediately served
as a check of the theory.

Let us now consider
further experimental verification
of the theory, —
but this time
the evidence did not precede
but was PREDICTED BY
the theory.

This was in connection with
the path of a ray of light
as it passes near a large mass
like the sun.

It will be remembered that
according to the Einstein theory
the presence of matter in space
makes the space non-Euclidean
and that the path of anything moving freely
(whether it be a planet
or a ray of light)
will be along a geodesic
in that space, and therefore
will be affected by the presence
of these obstacles in space.
Whereas,
according to classical physics,
the force of gravitation
could be exerted
only by one mass (say the sun)
upon another mass (say a planet),
but NOT upon a ray of light.

Here then was
a definite difference in viewpoint
between the two theories,
and the facts should
decide between them.
For this it was necessary
to observe
what happens to a ray of light
coming from a distant star
as it passes near the sun —
is it bent toward the sun,
as predicted by Einstein,
or does it continue on
in a straight line,
as required by classical physics? *
Now it is obviously impossible
to make this observation
under ordinary circumstances,
since we cannot look at a star
whose rays are passing near the sun,
on account of the brightness of the sun itself:
Not only would the star be invisible,
but the glare of the sun
would make it impossible
to look in that direction at all.

And so it was necessary
to wait for a total eclipse,

*If, however, light were considered
to be a stream of incandescent particles
instead of waves,
the sun WOULD have
a gravitational effect upon
a ray of light, even by classical theory,
BUT,
the AMOUNT of deflection
calculated even on this basis,
DOES NOT AGREE with experiment,
as we shall show later (see p. 287).

when the sun is up in the sky
but its glare is hidden by the moon,
so that the stars become
distinctly visible during the day.
Therefore, at the next total eclipse
astronomical expeditions were sent out
to those parts of the world
where the eclipse could be
advantageously observed,
and, —
since such an eclipse
lasts only a few seconds, —
they had to be prepared
to take photographs of the stars
rapidly and clearly,
so that afterwards,
upon developing the plates,
the positions of the stars
could be compared
with their positions in the sky
when the sun is NOT present.

The following diagram shows#

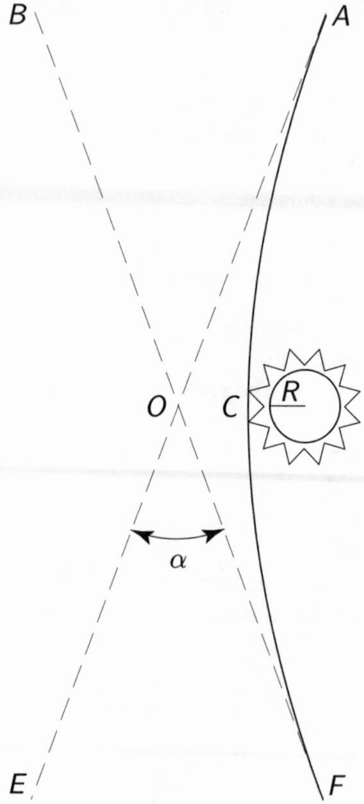

the path of a ray of light, *AOE*,
from a star, *A*,
when the sun is NOT
in that part of the sky.
And, also,
when the sun IS present,
and the ray is deflected
and becomes *ACF*,

#The deflection is greatly exaggerated
to make the tiny effect easy to see.

so that,

when viewed from F,

the star APPEARS to be at B.

Thus,

if such photographs

could be successfully obtained,

AND IF they showed

that all the stars

in that part of the sky near the sun

were really displaced (as from A to B)

AND IF

the MAGNITUDE of the displacements

agreed with the values

calculated by the theory,

then of course

this would constitute

very strong evidence in favor of

the Einstein theory.

Let us now determine

the magnitude of this displacement

as predicted by the Einstein theory:

We have seen (on page 233)

that

in the "Special Theory of Relativity,"

which applies in EUCLIDEAN space-time,

$$ds^2 = c^2dt^2 - (dx^2 + dy^2 + dz^2);$$

if we now divide this expression by dt^2,

we get

$$\left(\frac{ds}{dt}\right)^2 = c^2 - \left[\left(\frac{dx}{dt}\right)^2 + \left(\frac{dy}{dt}\right)^2 + \left(\frac{dz}{dt}\right)^2\right],$$

but

since $\dfrac{dx}{dt}, \dfrac{dy}{dt}, \dfrac{dz}{dt}$ are

281

the components of the velocity, v, of
a moving thing (see p. 263),
then obviously the
above quantity in brackets is v^2,
and the above equation becomes:

$$\left(\frac{ds}{dt}\right)^2 = c^2 - v^2.$$

Now when
the "moving thing" happens to be
a light-ray,
then $v = c$,
and we get, FOR LIGHT,

$$ds = 0.$$

But what about our
NON-EUCLIDEAN world,
containing matter?

It will be remembered (see p. 118)
that in studying a
non-Euclidean two-dimensional space
(namely, the surface of a sphere)
in a certain small region,
we were aided by
the Euclidean plane which
practically coincided with
the given surface in
that small region.
Using the same device for
space of higher dimensions,
we can,
in studying a region of
NON-Euclidean four-dimensional space-time,
such as our world is,
also utilize the
EUCLIDEAN 4-dimensional space-time which

practically coincides with it
in that small region.
And hence

$$ds = 0$$

will apply FOR LIGHT even
in our NON-EUCLIDEAN world.

And now,
using this result in (71),
together with the condition for
a geodesic, on page 261,
we shall obtain
THE PATH OF A RAY OF LIGHT.

XXXIII. DEFLECTION OF A RAY OF
LIGHT — *(Continued)*

In chapters XXIX and XXX we showed that
the condition for a geodesic
given on page 260
led to (74),
which, together with
the little g's on (71)
gave us the path of a planet, (84).

And now,
in order to find
the path of A RAY OF LIGHT,
we must add the further requirement:

$$ds = 0,$$

as we pointed out in Chapter XXXII.
Substututing $ds = 0$ in
the second equation of (84),

we get

$$h = \infty,$$

which changes the
first equation of (84) to

$$\frac{d^2u}{d\phi^2} + u = 3mu^2$$

which is the required
PATH OF A RAY OF LIGHT.

And this,
by integration*
gives, in rectangular coordinates,

$$x = R - \frac{m}{R} \cdot \frac{x^2 + 2y^2}{\sqrt{x^2 + y^2}}$$

for the equation of the curve on
page 280.

Now, since α (page 280) is
a very small angle,
the asymptotes of the curve may be
found by taking y very large by
comparison with x,
and so,
neglecting the x terms on the right
in the above formula,
it becomes

$$x = R - \frac{m}{R}(\pm 2y).$$

And,

*For details see Eddington, pp. 90–1;
or d'Inverno, pp. 199–201. [25]

using the familiar formula for
the angle between two lines, [26]

$$\tan \alpha = \frac{m_1 - m_2}{1 + m_1 m_2} \; ,$$

where α is the desired angle,
and m_1 and m_2 are
the slopes of the two lines,
we get

$$\tan \alpha = \frac{4Rm}{R^2 - 4m^2} \; ,$$

from which it is easy to find

$$\sin \alpha = \frac{4m}{R + 4m^2/R} \; .$$

And, α being small,
its value in radian measure is
equal, very nearly,[#] to $\sin \alpha$,
so that
we now have

$$\alpha = \frac{4m}{R + 4m^2/R} \tag{86}$$

Now,
what is the actual value of α
in the case under discussion,
in which

$$R = \text{the radius of the sun}$$

and

$$m = \text{its mass?}$$

[#] See any calculus textbook, e. g. [9],
   for the demonstration that
   $(\sin \theta)/\theta \to 1$ as $\theta \to 0$.
   Or, with a calculator
   (in radian mode),
   find values of $(\sin \theta)/\theta$
   for smaller and smaller θ:

$$\frac{\sin 0.5}{0.5} = 0.959, \; \frac{\sin 0.1}{0.1} = 0.998, \; \frac{\sin 0.02}{0.02} = 0.9999 \ldots$$

Since $R = 695,500$ kilometers,
and $m = 1.475$ kilometers*
$4m^2$ may be neglected by
comparison with R,
so that (86) reduces to the
very simple equation:

$$\alpha = \frac{4m}{R}$$

from which we easily get

$$\alpha = 1.75 \text{ seconds.}$$

In other words,
it was predicted by
the Einstein theory
that,
a ray of light passing near the sun
would be bent into a curve (ACF),
as shown in the figure on p. 280,
and that,
consequently
a star at A would
APPEAR to be at B,
a displacement of
an angle of 1.75 seconds!
If the reader will stop a moment
to consider
how small is an angle of
even one DEGREE,
and then consider that
one-sixtieth of that is
an angle of one MINUTE,
and again
one-sixtieth of that is

*See page 315. [27]

an angle of one SECOND,
she will realize how small is
a displacement of 1.75 seconds!

Furthermore,
according to the Newtonian theory,*
the displacement would be
only half of that!
And it is this TINY difference
that must distinguish
between the two theories.

After all the trouble that
the reader has been put to,
to find out the issue,
perhaps she is disappointed to learn
how small is the difference
between the predictions of
Newton and Einstein.
And perhaps she thinks that
a decision based on
so small a difference
can scarcely be relied upon!
But we wish to point out to her,
that,
far from losing her respect and faith
in scientific method,
she should,
ON THE CONTRARY,
be all the more filled with
ADMIRATION AND WONDER
to think that
experimental work in astronomy
IS SO ACCURATE
that

*See the footnote on p. 278. [28]

these small quantities* are measured
WITH PERFECT CONFIDENCE,
and they
DO distinguish
between the two theories and
DO decide in favor of the
Einstein theory,
as is shown by the
following figures:
The British expeditions, in 1919,
to Sobral and Principe,
gave for this displacement:

$$1.98'' \pm 0.12''$$

and

$$1.61'' \pm 0.30'',$$

respectively;
values which have since been
confirmed at other eclipses,
as, for example,
the results of Campbell and Trumpler,
who obtained,
using two different cameras,

$$1.72'' \pm 0.11'' \text{ and } 1.82'' \pm 0.15'',$$

in the 1922 expedition of the
Lick Observatory.

So that by now
all physicists agree that
the conclusions are
beyond question.

*See also the discrepancy in
the shift of the perihelion
of Mercury,
on page 275.

We cannot refrain,
in closing this chapter,
from reminding the reader that
1919 was right after World War I,
and that
Einstein was then classified as
a GERMAN scientist,
and yet,
the British scientists,
without any of the
stupid racial prejudices then
(and alas! still)
rampant in the world,
went to a great deal of trouble
to equip and send out expeditions
to test a theory by
an "enemy."

XXXIV. THE THIRD OF THE "CRUCIAL" PHENOMENA.

We have already seen that
two of the consequences from
the Einstein theory
were completely verified by
experiment:

(1) One, concerning the shift of
the perihelion of Mercury,
the experimental data for which
was known long before Einstein
BUT NEVER BEFORE EXPLAINED.
And it must be remembered
that the Einstein theory was

NOT expressly designed to
explain this shift,
but did it
QUITE INCIDENTALLY!

(2) The other, concerning the bending
of a ray of light as it
passes near the sun.
It was never suspected
before Einstein that
a ray of light when passing
near the sun
would be bent.*
It was for the first time
PREDICTED by this theory,
and, to everyone's surprise,
was actually verified
by experiment,
QUANTITATIVELY as well as
QUALITATIVELY (see Chap. XXXIII).

Now there was still another
consequence of this theory which
could be tested experimentally,
according to Einstein.
In order to appreciate it
we must say something about spectra.

Everyone probably knows that
if you hang a triangular glass prism
in the sunlight,
a band of different colors,
like a rainbow,
will appear on the wall where
the light strikes after it has
come through the prism.
The explanation of this phenomenon

*But, see the footnote on p. 278.

is quite simple,
as may be seen from the diagram:

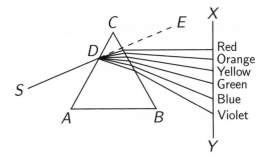

When a beam of white light, *SD*,
strikes the prism *ABC*,
it does NOT continue in
the SAME direction, *DE*,
but is bent.*
Furthermore,
if it is "composite" light,
like sunlight or any other white light,
which is composed of
light of different colors
(or different wavelengths),
each constituent
bends a different amount;

*This bending of a light ray
is called "refraction,"
and has nothing to do with
the bending discussed in Ch. XXXIII.
The reader may look up "refraction"
in any elementary physics textbook,
e. g. Holton and Brush, Chapter 23,
in Further Reading.

and when these constituents
reach the other side, *BC*, of the prism,
they are bent again,
as shown in the diagram on p. 291,
so that,
by the time they reach the wall, *XY*,
the colors are all separated out,
as shown,
the light of longest wavelength,
namely, red,
being deflected least.
Hence the rainbow-colored spectrum.

Now, obviously,
if the light from *S* is
"monochromatic,"
that is,
light of a SINGLE wavelength only,
instead of "composite,"
like sunlight,
we have instead of a "rainbow,"
a single bright line on *XY*,
having a DEFINITE position,
since the amount of bending,
as we said above,
depends upon the color or wavelength
of the light in question.
Now such monochromatic light
may be obtained from
the incandescent vapor
of a chemical element —
thus sodium, when heated,
burns with the light of
certain definite wavelengths,#
characteristic of sodium.

> #The principal wavelengths are two neighbors
> in the yellow part of the spectrum.

And similarly for other elements.
This is explained as follows:[#]
The atoms of each element
release and absorb light
of particular colors,
the characteristic wavelengths
of that element's spectrum.
This optical fingerprint
identifies an element uniquely.
Light is released (or absorbed)
by an atom's electrons
in falling (or rising)
from one energy level
to another.
The electrons are all the same,
but the energy levels are
particular to
the atoms of each element.
And so,
if you look at a spectrum
you can tell from the bright lines in it
just what substances
are present at S.

Each wavelength of light
has its own frequency,
a visual metronome
keeping the time.
According to Einstein,
the spectral lines of an element
thus may serve as
a measure for the "interval" ds,
between the beginning
and end of one wave,
and dt the time this takes,
the "period" of the light's oscillation.

[#] Prof. Lieber's description of the mechanism by which atoms emit characteristic colors of light was not clear, and so was rewritten. See note [29] on p. 344 for the original text.

Let an observer be
at rest relative to an atom
as it emits light.
Then, using space coordinates
such that

$$dr = d\theta = d\phi = 0,$$

that is,
locating the atom at the origin
of the observer's space coordinates,
equation (71) becomes

$$ds^2 = \gamma dt^2 \ \text{ or } \ ds = \sqrt{\gamma}\, dt,$$

where $\qquad \gamma = 1 - \dfrac{2m}{r}$ (see p. 253).

Now,
if an atom of, say, sodium,
emits light near the sun,
we should have to substitute
for m and r
the mass and radius of the sun;
and, similarly,
if an atom of the substance
emits light near the earth,
m and r would then have to be
the mass and radius of the earth,
and so on:
Thus γ DEPENDS upon
the location of the atom.
But since ds is
the space-time interval between
the beginning and end of a wave
as judged by an observer
at rest with respect to the atom,
ds is consequently
INDEPENDENT of the location
of the atom;
then, since $\qquad ds = \sqrt{\gamma}\, dt,$

294

obviously, *dt* would have to be
DEPENDENT UPON THE LOCATION.

Thus,
though sodium from a source
in a laboratory
gives rise to lines in
a definite part of the spectrum,
on the other hand,
sodium atoms in the sun
according to the above reasoning
reckon time at a different rate.
And so the light they emit will have
DIFFERENT frequencies
and hence the colors would be of
DIFFERENT wavelengths,
would then give bright lines in
DIFFERENT parts of the spectrum
from that ordinarily due to sodium.

And now let us see
HOW MUCH of a change in
the period of oscillation
is predicted by the Einstein theory
and whether it is borne out
by the facts:
If *dt* and *dt'* represent
the periods of oscillation near
the sun and the earth,
respectively,
then

$$ds = \sqrt{\gamma_{\text{sun}}}\; dt = \sqrt{\gamma_{\text{earth}}}\; dt'$$

or

$$\frac{dt}{dt'} = \frac{\sqrt{\gamma_{\text{earth}}}}{\sqrt{\gamma_{\text{sun}}}}.$$

295

Now γ_{earth} is very nearly equal to 1;
hence*

$$\frac{dt}{dt'} = \frac{1}{\sqrt{\gamma_{\text{sun}}}} = \frac{1}{\sqrt{1 - \dfrac{2m}{R}}}$$

$$= \frac{1}{1 - \dfrac{m}{R}} = 1 + \frac{m}{R}.$$

Or, using the values of
m and R given on page 286,
we get

$$\frac{dt}{dt'} = 1 + \frac{1.475}{695,500} = 1.00000212.$$

This result implies that
light emitted by a given element
should have a
slightly LOWER frequency
when it is near the sun than
when it is near the earth,
and hence a
slightly LONGER wavelength
and therefore
its lines should be
SHIFTED a little toward the
RED end of the spectrum (see p. 292).

This was a most unexpected result!
And since the amount of shift
was so slight,

*Neglecting higher powers of $\dfrac{m}{R}$
since $\dfrac{m}{R}$ is very small.

See note [7] to justify the approximation,

$$\frac{1}{\sqrt{1-x}} \approx 1 + \tfrac{1}{2}x \text{ if } x \ll 1.$$

it made the experimental verification
very difficult. [30]

For several years after
Einstein announced this result (1917)
experimental observations on this point
were doubtful,
and this caused many physicists
to doubt the validity of the
Einstein theory,
in spite of its other triumphs,
which we have already discussed.
BUT FINALLY, in 1927,
the very careful measurements
made by Evershed
definitely settled the issue
IN FAVOR OF THE EINSTEIN THEORY.

Furthermore,
similar experiments were performed
by W. S. Adams
on the star known as
the companion to Sirius,
which has a relatively
LARGE MASS and SMALL RADIUS, [31]
thus making the ratio

$$\frac{dt}{dt'} = 1 + \frac{m}{r}$$

much larger than
in the case of the sun (see p. 296)
and therefore easier to observe
experimentally.
Here too
the verdict was definitely
IN FAVOR OF THE EINSTEIN THEORY!

So that today

all physicists are agreed
that the Einstein theory
marks a definite step forward
for:

(1) IT EXPLAINED
    PREVIOUSLY KNOWN FACTS
    MORE ADEQUATELY THAN
    PREVIOUS THEORIES DID (see p. 103).

(2) IT EXPLAINED FACTS
    NOT EXPLAINED AT ALL
    BY PREVIOUS THEORIES
    such as:
    (a) The Michelson-Morley experiment,*
    (b) the shift in the perihelion of Mercury,†
    (c) the increase in mass of
        an electron when in motion.‡

(3) IT PREDICTED FACTS
    NOT PREVIOUSLY KNOWN AT ALL:
    (a) The bending of a light ray
        when passing near the sun.§
    (b) The shift of lines in
        the spectrum.**
    (c) The identity of mass and energy,
        which, in turn
        led to the ATOMIC BOMB!††

And all this
by using
VERY FEW
and

*See Part I, "The Special Theory."
†See Chapter XXXI.
‡See Chapter VIII.
§See Chapter XXXII.
**See pp. 289–297.
††See pp. 318–319.

VERY REASONABLE
hypotheses (see p. 97),
not in the slightest degree
"far-fetched" or "forced."

And what greater service
can any physical theory
render
than this!

We trust the reader
has been led by this little book
to have a sufficient insight
into the issues involved,
and to appreciate
the great breadth
and fundamental importance of
THE EINSTEIN THEORY OF RELATIVITY!

XXXV. SUMMARY.

I. In the SPECIAL Relativity Theory
   it was shown that
   two different observers,
   may, under certain
   SPECIAL conditions,
   study the universe from their
   different points of view
   and yet obtain
   the SAME LAWS and the SAME FACTS.

II. In the GENERAL Theory
    this democratic result was found to
    hold also for
    ANY two observers,
    without regard to the
    special conditions mentioned in I.

III. To accomplish this
Einstein introduced the
PRINCIPLE OF EQUIVALENCE,
by which
the idea of a FORCE OF GRAVITY
was replaced by
the idea of the
CURVATURE OF A SPACE.

IV. The study of this curvature
required the machinery of
the TENSOR CALCULUS,
by means of which
the CURVATURE TENSOR was derived.

V. This led immediately to
the NEW LAW OF GRAVITATION
which was tested by
the THREE CRUCIAL PHENOMENA
and found to work beautifully!

VI. And READ AGAIN
pages 298 and 299!

THE MORAL

THE MORAL

Since man has been
so successful in science,
can we not learn from
THE SCIENTIFIC WAY OF THINKING,
what the human mind is capable of,
and HOW it achieves SUCCESS:

I. There is NOTHING ABSOLUTE in science.
Absolute space and absolute time
have been shown to be myths.
We must replace these old ideas
by more human,
OBSERVATIONAL concepts.

II. But what we observe is
profoundly influenced by
the state of the observer,
and therefore
various observers get
widely different results —
even in their measurements of
time and length!

III. However,
in spite of these differences,
various observers may still
study the universe
WITH EQUAL RIGHT
AND EQUAL SUCCESS,
and CAN AGREE on
what are to be called
the LAWS of the universe.

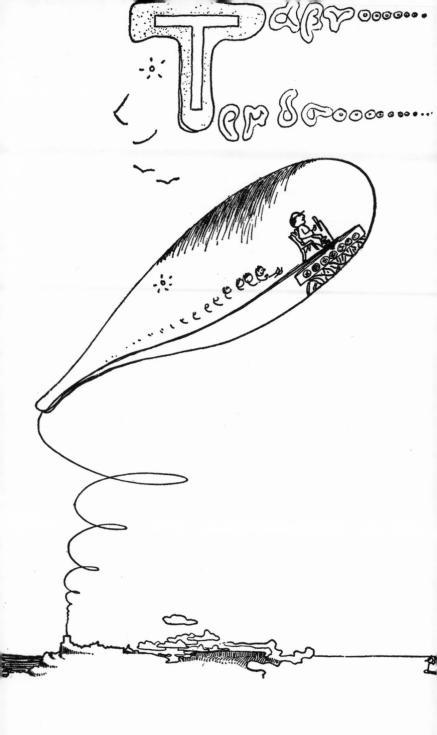

IV. To accomplish this we need
MORE MATHEMATICS
THAN EVER BEFORE,
MODERN, STREAMLINED, POWERFUL
MATHEMATICS.

V. Thus a combination of
PRACTICAL REALISM
(OBSERVATIONALISM)
and
IDEALISM (MATHEMATICS),
TOGETHER
have achieved SUCCESS.

VI. And,
knowing that the laws are
MAN-MADE,
we know that
they are subject to change
and we are thus
PREPARED FOR CHANGE.
But these changes in science
are NOT made WANTONLY,
BUT CAREFULLY AND CAUTIOUSLY
by the
BEST MINDS and HONEST HEARTS,
and not by any casual child who
thinks that
the world may be changed as easily
as rolling off a log.

WOULD YOU LIKE TO KNOW?

I. HOW THE EQUATIONS (20) ON PAGE 61
ARE DERIVED:

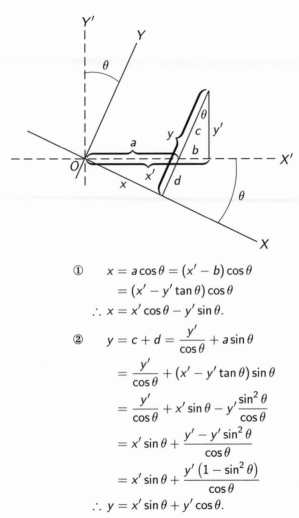

① $\quad x = a \cos \theta = (x' - b) \cos \theta$

$\qquad = (x' - y' \tan \theta) \cos \theta$

$\therefore x = x' \cos \theta - y' \sin \theta.$

② $\quad y = c + d = \dfrac{y'}{\cos \theta} + a \sin \theta$

$\qquad = \dfrac{y'}{\cos \theta} + (x' - y' \tan \theta) \sin \theta$

$\qquad = \dfrac{y'}{\cos \theta} + x' \sin \theta - y' \dfrac{\sin^2 \theta}{\cos \theta}$

$\qquad = x' \sin \theta + \dfrac{y' - y' \sin^2 \theta}{\cos \theta}$

$\qquad = x' \sin \theta + \dfrac{y' \left(1 - \sin^2 \theta\right)}{\cos \theta}$

$\therefore y = x' \sin \theta + y' \cos \theta.$

II. HOW THE FAMOUS MAXWELL EQUATIONS LOOK: [32]

$$\begin{cases} \dfrac{1}{c}\dfrac{\partial X}{\partial t} = \dfrac{\partial N}{\partial y} - \dfrac{\partial M}{\partial z} \\[2ex] \dfrac{1}{c}\dfrac{\partial Y}{\partial t} = \dfrac{\partial L}{\partial z} - \dfrac{\partial N}{\partial x} \\[2ex] \dfrac{1}{c}\dfrac{\partial Z}{\partial t} = \dfrac{\partial M}{\partial x} - \dfrac{\partial L}{\partial y} \end{cases}$$

$$\begin{cases} \dfrac{1}{c}\dfrac{\partial L}{\partial t} = \dfrac{\partial Y}{\partial z} - \dfrac{\partial Z}{\partial y} \\[2ex] \dfrac{1}{c}\dfrac{\partial M}{\partial t} = \dfrac{\partial Z}{\partial x} - \dfrac{\partial X}{\partial z} \\[2ex] \dfrac{1}{c}\dfrac{\partial N}{\partial t} = \dfrac{\partial X}{\partial y} - \dfrac{\partial Y}{\partial x} \end{cases}$$

X, Y, Z represent the components
of the ELECTRIC FIELD
at a point x, y, z in
an electromagnetic field,
at a given instant, t.

L, M, N represent the components
of the MAGNETIC FIELD
at the same point and
at the same instant.

III. HOW TO JUDGE
   WHETHER A SET OF QUANTITIES
   IS A TENSOR OR NOT: [33]

We may apply various criteria:

(1) See if it satisfies any of
    the definitions of tensors of
    various character and rank
    given in (16), (17), (18), etc.,
    or in (30), (31), etc.
    or in (32), etc.
Or
(2) See if is the
    sum, difference, or product
    of two tensors.
Or
(3) See if it satisfies
    the following theorem:
    A QUANTITY WHICH
    ON INNER MULTIPLICATION
    BY *ANY* COVARIANT VECTOR
    (OR *ANY* CONTRAVARIANT VECTOR)
    ALWAYS GIVES A TENSOR,
    IS ITSELF A TENSOR.
    This theorem may be
    quite easily proved
    as follows:
    Given that $X^{\alpha\beta\cdots}_{\gamma\delta\cdots} A_\alpha$ is known to be
    a contravariant vector,
    for any choice of
    the covariant vector A_α;
    To prove that $X^{\alpha\beta\cdots}_{\gamma\delta\cdots}$ is a tensor:
    Now since $X^{\alpha\beta\cdots}_{\gamma\delta\cdots} A_\alpha$ is
    a contravariant vector,
    it must obey (16), thus:

$$X'^{\alpha\beta\cdots}_{\gamma\delta\cdots} A'_\alpha = \frac{\partial x'_\beta}{\partial x_\nu} X^{\alpha\nu\cdots}_{\gamma\delta\cdots} A_\alpha;$$

312

but $\qquad A'_\alpha = \dfrac{\partial x_\mu}{\partial x'_\alpha} A_\mu$

or $\qquad A_\alpha = \dfrac{\partial x'_\mu}{\partial x_\alpha} A'_\mu,$

hence, by substitution,

$$X'^{\alpha\beta\cdots}_{\gamma\delta\cdots} A'_\alpha = \frac{\partial x'_\beta}{\partial x_\nu} X^{\alpha\nu\cdots}_{\gamma\delta\cdots} \frac{\partial x'_\mu}{\partial x_\alpha} A'_\mu$$

or# $\qquad \left[X'^{\alpha\beta\cdots}_{\gamma\delta\cdots} - \dfrac{\partial x'_\alpha}{\partial x_\mu} \dfrac{\partial x'_\beta}{\partial x_\nu} X^{\mu\nu\cdots}_{\gamma\delta\cdots} \right] A'_\alpha = 0.$

But A'_α does not have to be zero,
hence

$$X'^{\alpha\beta\cdots}_{\gamma\delta\cdots} = \frac{\partial x'_\alpha}{\partial x_\mu} \frac{\partial x'_\beta}{\partial x_\nu} X^{\mu\nu\cdots}_{\gamma\delta\cdots}$$

which satisfies (17),
thus proving that
$X^{\alpha\beta\cdots}_{\gamma\delta\cdots}$ must be a
CONTRAVARIANT TENSOR
OF RANK TWO
(and incidentally, that the other indices
must be contracted).

And similarly for other cases:
Thus if $X^{\alpha\beta\cdots}_{\gamma\delta\cdots} A^\gamma = B_{\mu\nu}$
then $X^{\alpha\beta\cdots}_{\gamma\delta\cdots}$ must be a tensor of
the form $C_{\gamma\mu\nu}$;
and if $X^{\alpha\beta\cdots}_{\gamma\delta\cdots} A_\alpha = C_{\mu\nu\rho}$,
then $X^{\alpha\beta\cdots}_{\gamma\delta\cdots}$ must be a tensor of
the form $D^\alpha_{\mu\nu\rho}$,
and so on.

Now let us show that
the set of little g's in (42)
is a tensor:

#We may swap the dummy indices α and μ
on the right-hand side.

313

Knowing that ds^2 is a
SCALAR —
i.e. A TENSOR OF RANK ZERO —
(see p. 128),
then
the right-hand member of (42) is also
A TENSOR OF RANK ZERO;
but dx_ν is, by (15) on p. 152,
A CONTRAVARIANT VECTOR,
hence,
by the theorem on page 312,
$g_{\mu\nu} \, dx_\mu$ must be
A COVARIANT TENSOR
OF RANK ONE.
And, again,
since dx_μ is
a contravariant vector,
then,
by the same theorem,
$g_{\mu\nu}$ must be
A COVARIANT TENSOR
OF RANK TWO,
and therefore
it is appropriate to write it
with TWO SUBscripts
as we have been doing
in anticipation of
this proof.

IV. WHY MASS CAN BE EXPRESSED IN KILOMETERS:

The reader may be surprised to
see the mass expressed in kilometers!
But it may seem more reasonable
from the following considerations:
In order to decide in what units
a quantity is expressed
we must consider its "dimensionality"
in terms of the fundamental units of
Mass, Length, and Time:
Thus the "dimensionality" of
a velocity is L/T;
the "dimensionality" of
an acceleration is L/T^2;
and so on.
Now, in Newtonian physics,
the force of an attraction which
the sun exerts upon the earth
being $F = kmm'/r^2$ (see p. 219),
where m is the mass of the sun,
m' the mass of the earth,
and r the distance between them;
and also, $F = m'a$,
a being the centripetal acceleration
of the earth toward the sun
(another one of the fundamental
laws of Newtonian mechanics);
hence

$$\frac{kmm'}{r^2} = m'a$$

or

$$m = \frac{1}{k}r^2 a.$$

Therefore,

the "dimensionality" of m is

$$L^2 \cdot \frac{L}{T^2} = \frac{L^3}{T^2}$$

since a constant, like k,
has no "dimensionality."
And now
if we take as a unit of time,
the time it takes light to go
a distance of one kilometer,
and call this unit
a "kilometer" of time
(thus 300,000 kilometers would
equal one second, since
light goes 300,000 kilometers in
one second),
then we may express
the "dimensionality" of m thus:
L^3/L^2 or simply L;
thus we may express
mass also in kilometers.
So far as considerations of
"dimensionality" are concerned,
the same result holds true also for
Einsteinian physics.
This idea of
"dimensionality"
is a very important tool
in scientific thinking. [34]

$$G_{11} = \{11,1\}\{11,1\} + \{11,2\}\{21,1\} +$$
$$\{11,3\}\{31,1\} + \{11,4\}\{41,1\} +$$
$$\{12,1\}\{11,2\} + \{12,2\}\{21,2\} +$$
$$\{12,3\}\{31,2\} + \{12,4\}\{41,2\} +$$
$$\{13,1\}\{11,3\} + \{13,2\}\{21,3\} +$$
$$\{13,3\}\{31,3\} + \{13,4\}\{41,3\} +$$
$$\{14,1\}\{11,4\} + \{14,2\}\{21,4\} +$$
$$\{14,3\}\{31,4\} + \{14,4\}\{41,4\}$$

$$- \frac{\partial}{\partial x_1}\{11,1\} - \frac{\partial}{\partial x_2}\{11,2\}$$

$$- \frac{\partial}{\partial x_3}\{11,3\} - \frac{\partial}{\partial x_4}\{11,4\}$$

$$+ \frac{\partial^2}{\partial x_1 \partial x_1} \log \sqrt{-g}$$

$$- \{11,1\} \frac{\partial}{\partial x_1} \log \sqrt{-g}$$

$$- \{11,2\} \frac{\partial}{\partial x_2} \log \sqrt{-g}$$

$$- \{11,3\} \frac{\partial}{\partial x_3} \log \sqrt{-g}$$

$$- \{11,4\} \frac{\partial}{\partial x_4} \log \sqrt{-g}$$

$$= 0.$$

If this mathematics BORES you
BE SURE TO READ
PAGES 318–323!

THE ATOMIC BOMB

We saw on p. 78 that the energy
which a body has when at rest, is:

$$E_o = mc^2.$$

Thus, the Theory of Relativity
tells us not only that
mass and energy are one and the same
but that, even though they are the same
still, what we consider to be
even a SMALL MASS is
ENORMOUS when translated into ENERGY terms,
so that a mass as tiny as an atom
has a tremendous amount of energy —
the multiplying factor being c^2,
the square of the velocity of light!
How to get at this great storehouse of
energy locked up in atoms
and use it to heat our homes,
to drive our cars and planes,
and so on and so on?
Now, so long as m is constant,
as for elastic collision,
E_o will also remain unchanged.
But, for inelastic collision,
m, and therefore E_o, will change;
and this is the situation when
AN ATOM IS SPLIT UP, for then
the sum of the masses of the parts
is LESS than the mass of the original atom.
Thus, if one could split atoms,
the resulting loss of mass would
release a tremendous amount of energy!
And so various methods were devised
by scientists like Meitner, Frisch,
Fermi, and others

318

to "bombard" atoms.
It was finally shown that when
Uranium atoms were bombarded with neutrons#
these atoms split up ("fission")
into two nearly equal parts,
whose combined mass is less than
the mass of the uranium atom itself,
this loss in mass being equivalent,
as the Einstein formula shows,
to a tremendous amount of energy,
thus released by the fission!
When Einstein warned President Roosevelt
that such experiments might lead to
the acquisition of terrific new sources of power
by the ENEMY of the human race,
the President naturally saw the importance
of having these experiments conducted
where there was some hope that
they would be used to END the war
and to PREVENT future wars
instead of by those who set out to
take over the earth for themselves alone!
Thus the ATOMIC BOMB
was born in the U.S.A.

And now that a practical method
of releasing this energy
has been developed,
the MORAL is obvious:
We MUST realize that it has become
too dangerous to fool around with
scientific GADGETS,
WITHOUT UNDERSTANDING
the MORALITY which is in

#S. Weinberg, *The Discovery of Subatomic Particles*
  (Cambridge: Cambridge Univ. Press, 2003);
  R. Rhodes, *The Making of the Atomic Bomb*,
  (New York: Simon & Schuster, 1995).

Science, Art, Mathematics —
SAM, for short.
These are NOT mere
idle words.
We must ROOT OUT the
FALSE AND DANGEROUS DOCTRINE
that SAM is amoral
and indifferent to
Good and Evil.
We must
SERIOUSLY EXAMINE SAM
FROM THIS VIEWPOINT.*

Religion has offered us
a Morality,
but many "wise guys" have
refused to take it
seriously,
and have distorted its
meaning!
And now, we are getting
ANOTHER CHANCE —
SAM is now also
warning us that
we MUST
UNDERSTAND the MORALITY which
HE is now offering us.
And he will not stand for
our failure to accept it,
by regarding him merely as
a source of gadgets!
Even using the atomic energy
for "peaceful" pursuits,

*See our book
 The Education of T. C. Mits
 for a further discussion
 of this vital point.

like heating the furnaces in
our homes,
IS NOT ENOUGH,
and will NOT satisfy SAM.
For he is desperately trying
to prevent us from
merely picking his pockets
to get at the gadgets in them,
and is begging us to see
the Good, the True, and
the Beautiful
which are in his mind and heart.
And, moreover,
he is giving
new and clear meanings to
these fine old ideas
which even the sceptical
"wise guys"
will find irresistible.
So
DO NOT BE AN
ANTI-SAMITE,
or SAM will get you
with his
atomic bombs,
his cyclotrons,
and all his new
whatnots.
He is so anxious to HELP us
if only we would listen
BEFORE IT IS TOO LATE!

FROM THE EDITORS

The text presented in this edition is nearly identical to the original (eleventh printing, 1961). The only changes, apart from those noted in the foreword, and a partial rewriting of pp. 293–296 (*cf.* note [29]) are these:

Mathematical infelicities or simple errors

p. 76, eq. (27): The equation "$mc/\sqrt{1-v^2/c^2} = mc(1-v^2/c^2)^{-\frac{1}{2}}$" replaces "$mc/\sqrt{1-v^2/c^2}$ or $mc(1-v^2/c^2)^{-\frac{1}{2}}$".

p. 80: The line "as electromagnetic waves," replaces "through an 'electromagnetic field,'".

p. 81: The phrases "electric field" and "magnetic field" replace "electric force" and "magnetic force", respectively. See also note [32].

pp. 251–252: The constant "C" replaces the constant "I".

p. 253, footnote: "We may for convenience let the constant equal $2m$ and write" replaces "We may then write."

p. 264: "$v = v_o$, a constant equal to the initial velocity" replaces "$v =$ a constant."

p. 264: "$s = v_o t + s_o$, $s_o =$ a constant equal to the initial position" replaces "$s = at + b$."

p. 264: "A STRAIGHT LINE (with slope v_o and intercept s_o)" replaces "A STRAIGHT LINE."

p. 272: The line "a is the semi-major-axis of the ellipse" has been deleted. Neither the variable a nor any other geometric factors of an ellipse had been introduced. Probably this was left over from an earlier draft.

p. 285: The expression "$\dfrac{4Rm}{R^2 - 4m^2}$" replaces "$\dfrac{4Rm}{4m^2 - R^2}$"; there was a sign error in the original denominator.

p. 285: The phrase "equal, very nearly" replaces the phrase "equal" (comparing the sine of a small angle with the angle itself).

p. 315: The conventional symbol for acceleration, "a", replaces "j".

Values for physical constants (e.g., the radius of the sun) have been updated throughout the text.

Footnotes

References to outmoded works have been replaced with citations to modern equivalents. For example, the last sentence of V., page 316, was truncated to remove a reference to an impossible to locate paper published by Prof. Lieber's institute. Instead, a recent reference was substituted in note [34]. When a new footnote has been introduced, or the footnote itself has been changed significantly, the footnote is marked with a sharp sign, thus: #. Examples occur on p. 280 and p. 313 (originally there were no footnotes) and on p. 285 (the original read, "For a proof, see any book on calculus, or look up a table of trigonometric functions.") People now use calculators, absent in Prof. Lieber's time; few people born after 1975 will have encountered trig tables.

Gender

In the early part of the twentieth century, few mathematicians were women: Prof. Lieber was unusual for her day. Her sensitivity to prejudice (*cf.* p. 289) and her subtle reminders of women in science (the Guide illustrated at the beginning of Part II is a woman, and Lise Meitner is the first nuclear scientist mentioned on p. 318) suggest that she would have welcomed the tendency of modern texts not to assume that every reader is male. Today, a postgraduate course in mathematics or in most fields of science will have about the same number of male and female students. Regrettably this is not yet true in physics. Textbooks in which all personal pronouns are masculine do not welcome half the population. In three places the gender of the reader could be excised gracefully. For example, on p. 237, the text originally reading "we hasten to console him" has been replaced by "we hasten to provide the consolation". About half the remaining masculine pronouns have been replaced with feminine equivalents. We think that Prof. Lieber would have written this way had it been the custom of her time.

p. 5

[1] The volume referred to in the footnotes on pp. 5, 13, 57, 81, and 96 is available in an inexpensive paperback edition: *The Principle of Relativity: A Collection of Original Papers on the Special and General Theory of Relativity*: Papers by A. Einstein, H. A. Lorentz, H. Weyl, H. Minkowski; Notes by A. Sommerfeld (New York: Dover, 1952), hereafter referred to as POR. This collection includes English translations of Einstein's first relativity paper, "On the Electrodynamics of Moving Bodies" (1905); the first derivation of $E = mc^2$, "Does the Inertia of a Body depend on its Energy Content?"; an early paper on the deflection of light in a gravitational field; and most of "The Foundation of the General Theory of Relativity," the 1916 paper that gave the world Einstein's theory of gravity.

p. 11

[2] What Prof. Lieber calls β is called γ by most authors;

$$\beta = \frac{c}{\sqrt{c^2 - v^2}} = \frac{1}{\sqrt{1 - (v/c)^2}} = \gamma.$$

Also, these same authors use β for the expression v/c.

p. 13

[3] Until 1959 the Lorentz contraction of a sphere into an ellipsoid was believed to appear just as Prof. Lieber describes. That year R. Penrose and J. Terrell independently showed that a photograph taken of a very rapidly moving sphere would show, surprisingly, a sphere: R. Penrose, "The Apparent Shape of a Relativistically Moving Sphere," *Proc. Camb. Phil. Soc.* 55 (1959): 137–139; J. Terrell, "Invisibility of the Lorentz Transformation," *Phys. Rev.* 116 (1959): 1041–1045.

p. 54

[4] The first description of a relativity principle appears to be Galileo's, in his famous book on the Copernican model of the solar system, *Dialogue Concerning the Two Chief World Systems*, trans. Stillman Drake (Berkeley: Univ. of California Press, 1953), 186–187. Galileo's relativity works fine for Newton's mechanics (developed only after Galileo's death) but not for Maxwell's electromagnetism.

p. 57

[5] Hermann Minkowski was one of Einstein's professors, at first incredulous that the student he once called a "lazy dog" had discovered relativity. Describing relativity at a 1908 assembly of scientists and physicians in Cologne, Germany, Minkowski said famously

> The views of space and time which I wish to lay before you have sprung from the soil of experimental physics, and therein lies their strength. They are radical. Henceforth space by itself, and time by itself, are doomed to fade away, and only a kind of union of the two will preserve an independent reality.

327

(H. Minkowski, "Space and Time," in POR.) A few months later, Minkowski died from a botched emergency appendectomy. At first Einstein was very skeptical of Minkowski's geometric approach to relativity. He soon realized, however, that geometry would allow him to find the general theory.

p. 64
[6] The replacement $t = i\tau$ was popular until about forty years ago. With the first edition of Taylor and Wheeler's textbook (see Further Reading), this convention, and the use of the trigonometric functions sine and cosine for Lorentz transformations as on p. 65, began to fall out of favor. Instead, one retains the time t and introduces the *hyperbolic* trigonometric functions *sinh* and *cosh* (rhyme with "pinch" and "gosh," respectively). In place of the familiar identity

$$\sin^2 \theta + \cos^2 \theta = 1$$

one has

$$\cosh^2 \theta - \sinh^2 \theta = 1.$$

Recalling Euler's identity

$$e^{ix} = \cos x + i \sin x$$

it follows that

$$\cos x = \frac{e^{ix} + e^{-ix}}{2} \qquad \sin x = \frac{e^{ix} - e^{-ix}}{2i},$$

which invites the definitions

$$\cosh x = \frac{e^x + e^{-x}}{2} \qquad \sinh x = \frac{e^x - e^{-x}}{2}.$$

p. 76
[7] Everyone knows the identity

$$(a + b)^2 = a^2 + 2ab + b^2.$$

Let $a = 1$, and let b equal a small number, for example, $\frac{1}{100}$. Then

$$(1 + \tfrac{1}{100})^2 = 1.01^2 = 1.0201.$$

This number is not very different from 1.02. So, using $\doteq$ to indicate near equality, we can write

$$(1 + \tfrac{1}{100})^2 \doteq 1 + 2 \times (0.01).$$

That is, if b is much smaller than a (written $b \ll a$), we can to a reasonable approximation write

$$(a + b)^2 \doteq a^2 + 2ab,$$

that is, we can simply drop the term b^2; if $b \ll a$, then b^2 is negligible in comparison with a^2. In a similar manner we can say, provided $x \ll 1$, that

$$\frac{1}{1 - x} \doteq 1 + x.$$

To show this, multiply each side by $1 - x$; then

$$1 \doteq (1 - x)(1 + x) = 1 - x^2 \doteq 1$$

as claimed. In the same way, it is easy to see (recall $\sqrt{a} = a^{1/2}$) that

$$(1 + x)^{1/2} \doteq 1 + \tfrac{1}{2}x$$

provided $x \ll 1$; just square both sides to see that it is true. Consequently, ignoring terms in x of a power higher than the first,

$$\frac{1}{\sqrt{1 - x}} \doteq 1 + \tfrac{1}{2}x,$$

which is just Eq. (28). Prof. Lieber is using a more general identity;

$$(a + b)^n = a^n + na^{n-1}b + \frac{n(n - 1)}{2!}a^{n-2}b^2 + \ldots$$

where $p! = p \times (p - 1) \times (p - 2) \times \cdots \times 1$, and by convention $0! = 1$. Substituting $a = 1$, $b = -(v/c)^2$, and $n = -\tfrac{1}{2}$ gives

$$\begin{aligned}
(1 - (v/c)^2)^{-1/2} &= 1^{-1/2} + \left(-\tfrac{1}{2}\right)1^{-3/2} \times \left(-(v/c)^2\right) \\
&\quad + \left(-\tfrac{1}{2} \times -\tfrac{3}{2}\right) \times \tfrac{1}{2!} \times \left(-(v/c)^2\right)^2 + \ldots \\
&= 1 + \tfrac{1}{2}(v/c)^2 + \tfrac{3}{8}(v/c)^4 + \ldots,
\end{aligned}$$

which is exactly the same as the terms in the parentheses on the bottom of p. 76. Dropping terms in (v/c) higher than the second gives Eq. (28).

p. 79

[8] Rather than say that the mass of a rapidly moving object increases with speed, it is now standard to regard the mass as an invariant (as Prof. Lieber describes Einstein's preference on p. 80), and to say that the object's *inertia* (its resistance to acceleration) increases by the factor β as its speed increases. In the limit as the speed approaches c, the inertia approaches ∞. Simply put, an object whose mass is not zero would require an infinite amount of energy to reach the speed of light. The famous equation is better written as

$$E = \beta mc^2,$$

which reduces to the familiar version if the object is at rest, because $\beta = 1$ if $v = 0$. The first experimental evidence that the inertia of an object increased with speed was obtained by Walter Kaufmann in 1901, before Einstein derived the result. He seems not to have known of Kaufmann's results before finding the equivalence between mass and energy. After Einstein's work appeared Kaufmann repeated his experiments. Again he obtained an increase of inertia with speed, but the data did not seem in accord with Einstein's formula. Many believed Kaufmann; Einstein, Planck, and a few others judged the data unreliable. Definitive experiments were performed during 1914–1916, verifying Einstein's equation. See S. Goldberg, *Understanding Relativity: Origin and Impact of a Scientific Revolution* (Boston: Birkhäuser, 1984), 134–149, and A. Pais, *Subtle is the Lord: The*

Science and the Life of Albert Einstein (New York: Oxford Univ. Press, 1982), 156–159.

p. 144

[9] Most calculus textbooks are thick, costly, and forbidding. Here are two which are thin, inexpensive, and very informal: S. P. Thompson and Martin Gardner, *Calculus Made Easy* (New York: St. Martin's Press, 1998) and Michael Spivak, *The Hitchhiker's Guide to Calculus* (Washington, DC: Mathematical Association of America, 1995).

[10] Suppose we have a system of two linear equations in two dimensions, of the form

$$ax + by = x'$$
$$cx + dy = y'$$

and we'd like to solve these for x and y. A unique solution exists if and only if the quantity $ad - bc$, the *determinant*, is not equal to zero. Many algebra and trigonometry texts discuss determinants. An accessible introduction to matrices and determinants may be found in D. E. Littlewood, *The Skeleton Key of Mathematics* (New York: Dover, 2002), chap. XXI. Prof. Lieber discusses determinants further on p. 193 and p. 195. See also the footnote on p. 239.

p. 147

[11] Partial differentiation, described more fully by Prof. Lieber on p. 204, is an operation from advanced calculus. Here is the idea in a nutshell. For an ordinary function $f(x)$, define the derivative of f at a point $(a, y = f(a))$ with the familiar rule

$$\frac{df}{dx}\bigg|_{x=a} = \lim_{h \to 0} \frac{f(a + h) - f(a)}{h}.$$

Imagine a function $f(x, y)$ which depends on two independent variables, x and y. Each of these variables may be changed while keeping the other fixed in value. To graph the function requires *three* coordinates: x, y, and a third, $z = f(x, y)$; typically z is going to be a *surface*, a set of hills and valleys, maybe. Pick some value of y, say $y = b$. The set of all points satisfying $y = b$ is a plane parallel to the xz-plane. The intersection of this plane with the surface is a curve, given by

$$z = f(x, b).$$

Now, it is easy to take the derivative of z with respect to x at a point $(x = a, y = b, z = f(a, b))$. In fact, there is no need to fix the value of y; it can be anything you like. To distinguish this derivative (which is the slope of the curve $z = f(x, y)$ with respect to x) from an ordinary derivative of a function $f(x)$ of a single variable, the symbol "∂" is used:

$$\frac{\partial f}{\partial x}\bigg|_{x=a, y} = \lim_{h \to 0} \frac{f(a + h, y) - f(a, y)}{h}.$$

To perform this operation, merely treat all the other variables (the ones that you are *not* taking the derivative with respect to) as if they

330

were constants. For example, consider $f(x, y) = x^2 + xy + y^3$. Then

$$\frac{\partial f}{\partial x} = 2x + y \qquad \frac{\partial f}{\partial y} = x + 3y^2.$$

That's all there is to it.

p. 178

[12] For an accessible, inexpensive book on tensor calculus, see D. F. Lawden, *Introduction to Tensor Calculus, Relativity and Cosmology* (New York: Dover, 1982). Lawden also uses $\tau = it$.

p. 191

[13] Prof. Lieber devotes the next few pages to constructing the covariant metric tensor $g^{\mu\nu}$. She does not mention that this tensor is the algebraic inverse of the contravariant metric tensor $g_{\mu\nu}$; that is, the multiplication of these two matrices produces the identity matrix with 1's along the diagonal and 0's elsewhere. In tensor notation, (summation on λ!)

$$g^{\mu\lambda} g_{\lambda\nu} = \delta^\mu_\nu$$

where δ^μ_ν is the *Kronecker delta*, the identity matrix;

$$\delta^\mu_\nu = \begin{cases} 1, & \text{if } \mu = \nu; \\ 0, & \text{if } \mu \neq \nu \end{cases} = \begin{pmatrix} 1 & 0 & 0 \\ 0 & 1 & 0 \\ 0 & 0 & 1 \end{pmatrix} \quad \text{for a } 3 \times 3 \text{ matrix.}$$

p. 196

[14] The notation $\{\mu\nu, \lambda\}$ given by equation (46) is outmoded. The Christoffel symbol is usually written as $\Gamma^\lambda_{\mu\nu}$. A second form of Christoffel symbol, hinted at by Prof. Lieber in the footnote on p. 196, is written old-style as $[\mu\nu, \lambda]$ and in new notation, $\Gamma_{\lambda\mu\nu}$.

p. 206

[15] The Riemann-Christoffel curvature tensor, denoted $B^\alpha_{\sigma\tau\rho}$ by Prof. Lieber, is today more frequently denoted $R^\alpha_{\sigma\rho\tau} = -R^\alpha_{\sigma\tau\rho}$.

p. 215

[16] The Einstein tensor $G_{\sigma\tau}$ is given by the expression

$$G_{\sigma\tau} = B^\alpha_{\sigma\tau\alpha} - \tfrac{1}{2} g_{\sigma\tau} g^{\lambda\mu} B^\alpha_{\lambda\mu\alpha}.$$

The *trace* (German *Spur*) of a matrix is the sum of its diagonal elements. This operation is accomplished for a tensor of rank n through "contraction": set one subscript equal to one superscript, and do the implied addition (*cf.* p. 182). The contraction

$$B^\alpha_{\sigma\tau\alpha} \equiv B_{\sigma\tau} = B^1_{\sigma\tau 1} + B^2_{\sigma\tau 2} + B^3_{\sigma\tau 3} + B^4_{\sigma\tau 4}$$

is called the *Ricci tensor*; a further contraction

$$g^{\sigma\tau} B_{\sigma\tau} = B^\tau_\tau \equiv B$$

is called the *Ricci scalar*. That is, the Einstein tensor may be written

$$G_{\mu\nu} = B_{\mu\nu} - \tfrac{1}{2} g_{\mu\nu} B.$$

The full Einstein equation is (in units where $c = 1$)

$$G_{\mu\nu} = -8\pi k T_{\mu\nu},$$

where k is the constant in Newton's Law of Gravitation (see p. 219), and the tensor $T_{\mu\nu}$ is the *energy-momentum tensor* of the mass in some region of space. In empty space (for example, near a star) we can say $T_{\mu\nu} = 0$. Then the Einstein equation says

$$G_{\mu\nu} = 0.$$

A significant simplification occurs if we take the trace of this equation. Because (see note [13])

$$g^{\mu\lambda}g_{\lambda\nu} = \delta^{\mu}_{\nu},$$

it follows that

$$g^{\mu\lambda}g_{\lambda\mu} = \delta^{\mu}_{\mu} = 4,$$

so that, using the definition of the Einstein tensor,

$$0 = g^{\mu\nu}G_{\mu\nu} = B - \tfrac{1}{2}\delta^{\mu}_{\mu}B = B - 2B = -B;$$

in empty space, the Ricci scalar equals zero, and consequently the Einstein tensor $G_{\mu\nu}$ is exactly equal to the Ricci tensor $B_{\mu\nu}$. It is this equation,

$$B_{\mu\nu} = G_{\mu\nu} = 0,$$

that the rest of the book discusses.

p. 223
[17] The symbol "∇" is most often called *del*. It was introduced by the Irish mathematician William Rowan Hamilton, who named it *nabla* after the Hebrew word for "harp," which he thought ∇ resembled. (Hamilton had learned more than a dozen languages by the age of 12, including Hebrew.) Today the symbol ∇^2 is usually pronounced "del squared".

p. 233
[18] Prof. Lieber devotes the next twenty-two pages to deriving the famous *Schwarzschild metric*. In plain English, Einstein's equation for gravity says

(curvature) = (constant) × (energy-momentum tensor).

(In empty space, the right-hand side is zero.) The catch is that gravity, manifest in the curvature, is *itself* a form of energy. This makes the Einstein equations *non-linear*. *Linear* differential equations are solved with standard methods, but non-linear differential equations are very difficult to solve, and usually lack exact solutions. Einstein thus resorted to an approximation to determine his theory's answers to the three classic tests (perihelion of Mercury; deflection of light; gravitational red shift). A few months later, to Einstein's surprise and delight, an *exact* solution was found by the astronomer Karl Schwarzschild (then a German soldier in the Great War). Schwarzschild's solution, describing the geometry around a static, spherical mass like the sun, gives exactly the same answers to the three tests as Einstein's approximation. Schwarzschild died soon after finding his solution (May 1916) from an illness contracted during the war. His son Martin (1912–1997) became a distinguished

332

astrophysicist at Princeton. Schwarzschild's solution predicts the existence of *black holes*; the boundary beyond which an object will inexorably fall into the black hole is called the *event horizon*; its distance from the center of the black hole is called the *Schwarzschild radius*.

p. 239
[19] The formula for the determinant of a 2 × 2 matrix is given in note [10]. Analogous formulas for determinants of larger matrices are cumbersome. A better method is to calculate determinants as a sum of smaller determinants. Define the *minor* of a matrix element as the determinant of the smaller matrix resulting from striking out the row and column containing that element. The *cofactor* of a matrix element is that element's minor times $(-1)^{i+j}$, where i = the row number and j = the column number. Like riding a bicycle, this is easier to show than to explain. For Prof. Lieber's matrix on p. 195,

$$\begin{vmatrix} \cancel{5} & \cancel{2} & \cancel{3} \\ 4 & 1 & 0 \\ 6 & 8 & 7 \end{vmatrix} \quad \text{The cofactor of 5} = (+1) \times \begin{vmatrix} 1 & 0 \\ 8 & 7 \end{vmatrix} = 7.$$

$$\begin{vmatrix} 5 & 2 & 3 \\ \cancel{4} & \cancel{1} & \cancel{0} \\ 6 & 8 & 7 \end{vmatrix} \quad \text{The cofactor of 4} = (-1) \times \begin{vmatrix} 2 & 3 \\ 8 & 7 \end{vmatrix} = 10.$$

$$\begin{vmatrix} 5 & 2 & 3 \\ 4 & 1 & 0 \\ \cancel{6} & \cancel{8} & \cancel{7} \end{vmatrix} \quad \text{The cofactor of 6} = (+1) \times \begin{vmatrix} 2 & 3 \\ 1 & 0 \end{vmatrix} = -3.$$

The matrix element 5 appears in the first row and first column, so its minor is multiplied by $(-1)^{1+1} = (-1)^2 = 1$; the element 6 is in the third row and first column, so its minor is multiplied by $(-1)^{3+1} = (-1)^4 = 1$.

To find det M, the determinant of Prof. Lieber's matrix, choose a row or a column. Here we've chosen the first column. The determinant of M is given very simply by this equation:

$$\det M = 5 \times \text{cof}(5) + 4 \times \text{cof}(4) + 6 \times \text{cof}(6)$$
$$= 5 \times 7 + 4 \times 10 + 6 \times (-3) = 57.$$

We can do this with any row or any column, no matter how large the matrix. The matrix has to be *square*, and we have to stay on the row or column we choose. Suppose you have, as on p. 239, a 4 × 4 matrix. Let's go across the top row;

$$\det g_{\mu\nu} = g = (-e^{\lambda}) \times \text{cof}(-e^{\lambda}) + 0 \times \text{cof}(0)$$
$$+ 0 \times \text{cof}(0) + 0 \times \text{cof}(0)$$
$$= (-e^{\lambda}) \times \text{cof}(-e^{\lambda})$$

and we don't even have to bother with calculating any other cofactors besides the first, because the other elements along the top row are

333

all zero! The cofactor of $-e^\lambda$ is $(+1)$ times the 3×3 determinant

$$\begin{vmatrix} -x_1^2 & 0 & 0 \\ 0 & -x_1^2 \sin^2 x_2 & 0 \\ 0 & 0 & e^\nu \end{vmatrix}.$$

We know how to find this 3×3 determinant. Pick a row or column, say the top row. Observe that two of the row elements are 0, so the determinant reduces to a single term: $-x_1^2$ times its cofactor, $(+1)$ times a 2×2 determinant, equal to the product of the two diagonal terms. So g is nothing but the product of all four non-zero elements along the diagonal. This is a general rule for a diagonal matrix M (all the numbers not on the main diagonal are zero): $\det M =$ the product of the diagonal elements.

p. 242

[20] The reduction of the ten Einstein equations $G_{\mu\nu} = 0$ to six is important historically and physically. To obtain n unknown functions from solving differential equations requires n independent equations, no more and no less. The Einstein equations were supposed to determine uniquely the ten components of the metric tensor $g_{\mu\nu}$. Since energy and momentum are *conserved*, it turns out that the four covariant derivatives of the energy-momentum tensor must equal zero. The Einstein tensor is proportional to the energy-momentum tensor, so its four covariant derivatives must also equal zero. That leaves six independent components of $G_{\mu\nu} = 0$ to be solved. Einstein had the right equations in 1913, but was perplexed by two problems. First, it was not obvious to him that the covariant derivatives of $G_{\mu\nu}$ had to be zero. That they must, follows from identities worked out by the Italian geometer Luigi Bianchi many years earlier (but which were unknown to Einstein in 1913). Second, even if the equations could be reduced from ten to six, there were apparently still ten components of the metric tensor $g_{\mu\nu}$ to be obtained, now from only six equations. In the fall of 1915, he realized that he had a four-fold freedom (nowadays called *gauge invariance*, also found in Maxwell's theory of electromagnetism) in the choice of the metric tensor, reducing the ten g's to only six independent components. Hence there were precisely as many equations, six, as components of the metric tensor to be found. For more on the development of general relativity, see A. Pais, op. cit. (note [8]). For more about Luigi Bianchi, see the Wikipedia site http://en.wikipedia.org/wiki/Luigi_Bianchi. It seems that Bianchi *rediscovered* the identities; they had been found twenty years earlier by another famous Italian geometer, Tullio Ricci, whose name graces the tensor in the note to p. 215.

p. 257

[21] For an accessible introduction to the calculus of variations in physics, see Don S. Lemons, *Perfect Form*, (Princeton: Princeton U. Press, 1997). Paul J. Nahin provides both an engaging history and a friendly tutorial of the mathematics of minimal principles in *When Least Is Best* (Princeton: Princeton Univ. Press, 2003); see chap. 6, "Beyond Calculus," pp. 200–271, for Prof. Nahin's discussion of the calculus of variations.

[22] Eddington's book, once inexpensive to obtain, is out of print and commands a king's ransom. For convenience, here is a sketch of his arguments, as cited by Prof. Lieber. His now standard calculation can be found in most textbooks on general relativity, e. g., d'Inverno's. (See Further Reading.)

Equation (82) reads

$$\frac{d^2\phi}{ds^2} + \frac{2}{r}\frac{dr}{ds}\frac{d\phi}{ds} = 0. \tag{82}$$

Consider the expression

$$r^2\frac{d\phi}{ds} = h \, ; \tag{A}$$

then by differentiation with respect to s

$$\frac{dh}{ds} = 2r\frac{dr}{ds}\frac{d\phi}{ds} + r^2\frac{d^2\phi}{ds^2} = r^2\left(\frac{d^2\phi}{ds^2} + \frac{2}{r}\frac{dr}{ds}\frac{d\phi}{ds}\right) = 0.$$

So, from Eq. (82) it follows that $h = r^2\dfrac{d\phi}{ds}$ is a constant. We get a second constant from Eq. (83):

$$\frac{d^2t}{ds^2} + \nu'\frac{dr}{ds}\frac{dt}{ds} = 0 = \frac{d^2t}{ds^2} + \frac{d\nu}{ds}\frac{dt}{ds} \tag{83}$$

(recall that $\dfrac{d\nu}{ds} = \dfrac{d\nu}{dr}\dfrac{dr}{ds} = \nu'\dfrac{dr}{ds}$). Consider the expression

$$b = e^\nu\frac{dt}{ds}.$$

Differentiate this expression with respect to s:

$$\frac{db}{ds} = e^\nu\frac{d\nu}{ds}\frac{dt}{ds} + e^\nu\frac{d^2t}{ds^2} = e^\nu\left(\frac{d^2t}{ds^2} + \frac{d\nu}{ds}\frac{dt}{ds}\right) = 0.$$

That is, from Eq. (83) if follows that $b = e^\nu\dfrac{dt}{ds}$ is a constant, or

$$\frac{dt}{ds} = be^{-\nu} = \frac{b}{\gamma} \tag{B}$$

(see p. 252). We leave alone Eq. (81), instead returning to Eq. (71) (p. 255), recalling (p. 268) that we may without loss of generality restrict the value of θ to be equal always to $\pi/2$ for objects following a geodesic near the sun; the path lies in a plane. Making this substitution in Eq. (71) leads to

$$ds^2 = -\frac{1}{\gamma}\,dr^2 - r^2\,d\phi^2 + \gamma\,dt^2.$$

Multiply both sides by $-\gamma/ds^2$, substitute in for $d\phi/ds$ (from (A)) and for dt/ds (from (B)) to obtain

$$-\gamma = \left(\frac{dr}{ds}\right)^2 + \frac{\gamma h^2}{r^2} - b^2.$$

From (A) it follows that $ds = (r^2/h)d\phi$, so substituting this into the last equation gives

$$-\gamma = \left(\frac{h}{r^2} \frac{dr}{d\phi} \right)^2 + \frac{\gamma h^2}{r^2} - b^2,$$

which, using $\gamma = 1 - (2m/r)$ (see p. 253), may be rewritten as

$$\left(\frac{h}{r^2} \frac{dr}{d\phi} \right)^2 + \frac{h^2}{r^2} = b^2 - 1 + \frac{2m}{r} + \frac{2mh^2}{r^3}.$$

Now let $u = \frac{1}{r}$, so $r = \frac{1}{u}$. Then

$$\frac{dr}{d\phi} = -\frac{1}{u^2} \frac{du}{d\phi}$$

and the last equation may be written as

$$\left(\frac{du}{d\phi} \right)^2 + u^2 = \frac{b^2 - 1}{h^2} + \frac{2mu}{h^2} + 2mu^3.$$

Differentiate with respect to ϕ, and obtain

$$2 \left(\frac{du}{d\phi} \right) \frac{d^2u}{d\phi^2} + 2u \frac{du}{d\phi} = \frac{2m}{h^2} \frac{du}{d\phi} + 6mu^2 \frac{du}{d\phi}.$$

This may be simplified by dividing by $2du/d\phi$ to obtain finally

$$\frac{d^2u}{d\phi^2} + u = \frac{m}{h^2} + 3mu^2, \qquad (C)$$

which, together with (A), is Prof. Lieber's Eq. (84).

p. 271

[23] Prof. Lieber wants to compare Einstein's relativistic pair of equations (84) on p. 270 with the Newtonian pair (85) on p. 271. The idea is that Newton's equations are an approximation to Einstein's, the latter approaching the former as $(v/c) \to 0$ (where v is an object's speed.) This is complicated a little by the fact that h in one equation is not quite the same quantity as in the other. We have to be careful.

The two equations in (85) follow from the calculation for the path of a planet in Newtonian physics. Newton's Second Law is $(\alpha = 1, 2)$

$$F_\alpha = m_p a_\alpha,$$

where F_α is the net force acting on a planet of mass m_p, and a_α is the planet's acceleration. Use polar coordinates: $x_1 = r$; $x_2 = \phi$. The sun with mass m is placed at the origin. The planet's position is specified by its distance r from the sun and its orientation ϕ with respect to the usual x-axis. The size of the force acting on the planet is given by Newton's Law of Gravitation,

$$F = \frac{kmm_p}{r^2},$$

where k is Newton's gravitational constant (see p. 219). This is an attractive force, toward the sun, directed along the line connecting

336

the planet and the sun. There is no force in the ϕ direction. Using the expressions for the radial (r) and tangential (ϕ) components of acceleration, Newton's Second Law gives, for $\alpha = 1$,

$$-\frac{kmm_p}{r^2} = m_p\left(\frac{d^2r}{dt^2} - r\left(\frac{d\phi}{dt}\right)^2\right),$$

and for $\alpha = 2$,

$$0 = m_p\left(2\frac{dr}{dt}\frac{d\phi}{dt} + r\frac{d^2\phi}{dt^2}\right).$$

This equation can be written as

$$m_p\frac{d}{dt}\left(r^2\frac{d\phi}{dt}\right) = 0,$$

so that

$$r^2\frac{d\phi}{dt} = h, \tag{D}$$

where h is a constant equal to L/m_p; L is a constant called the *angular momentum* (see note [24]). This equation (D) is the second of the pair of equations in (85). The first of the pair (85) comes from the radial component of Newton's Law ($\alpha = 1$). Divide this equation by m_p, convert the units of time and kilograms to meters by replacing km with mc^2 and t by t/c, and substitute $d\phi/dt = h/r^2$. Finally, divide by c^2. What results is

$$\frac{d^2r}{dt^2} = -\frac{m}{r^2} + \frac{h^2}{r^3}. \tag{E}$$

This is, in relativistic units, the radial component of Newton's Law of Gravitation. To make a comparison with the relativistic pair of equations (84) (p. 270), we seek an equation for $u(\phi)$, not $r(t)$, where $u = 1/r$. By the Chain Rule,

$$\frac{dr}{dt} = \frac{d(1/u)}{d\phi}\frac{d\phi}{dt} = -\frac{1}{u^2}\frac{du}{d\phi}\cdot\frac{h}{r^2}$$

$$= -h\frac{du}{d\phi}$$

$$\frac{d^2r}{dt^2} = \frac{d}{d\phi}\left(-h\frac{du}{d\phi}\right)\frac{d\phi}{dt} = -h\frac{d^2u}{d\phi^2}\cdot\frac{h}{r^2}$$

$$= -h^2u^2\frac{d^2u}{d\phi^2}.$$

Substitute this expression into (E) and write $(1/r) = u$ to get

$$-h^2u^2\frac{d^2u}{d\phi^2} = -mu^2 + h^2u^3. \tag{F}$$

Divide by $-h^2u^2$ and rearrange terms to obtain

$$\frac{d^2u}{d\phi^2} + u = \frac{m}{h^2}, \tag{G}$$

which is called *Binet's equation*, and is the first of the pair of Prof. Lieber's equations (85). The solution of Binet's equation is that the path of a planet must be a *conic section*: circle, ellipse, parabola, or hyperbola. This is a standard calculation to be found in any upper-level mechanics text, e. g., J. R. Taylor, *Classical Mechanics*, (Herndon, VA: University Science Books, 2005), chapter 8. Richard P. Feynman provides a characteristically unconventional and elegant demonstration that Newton's inverse-square law for gravity leads to elliptical orbits in David L. and Judith R. Goodstein, *Feynman's Lost Lecture: The Motion of Planets Around the Sun*, with CD (New York: W. W. Norton, 2000); without CD (New York: Vintage, 1997). Feynman makes use of *geometry*, not calculus! (The solution to (G), Binet's equation, is equation (J), below; set $km/c^2 = m$ to convert (J) to relativistic units.)

pp. 272
[24] In discussing the perihelion advance of Mercury, Prof. Lieber again refers the reader to Eddington's book. For convenience here are the details (Eddington, pp. 88–90; d'Inverno, pp. 195–198).

Imagine a massive object (the planet Mercury, or something else) in orbit around the sun. The relevant equation is (C) above; but use the conventional mass (kilograms) by replacing m with km/c^2 to obtain:

$$\frac{d^2u}{d\phi^2} + u = \frac{km}{c^2h^2} + \frac{3kmu^2}{c^2}. \tag{H}$$

The analogous Newtonian equation is equation (G), but which now has to be in conventional units. Unfortunately, h in the relativistic equation (H) is not exactly the same as in the Newtonian equation (G). To keep these distinct, replace h in equation (G) by L/m_pc. L is the orbiting object's angular momentum, equal to m_pvr, where m_p is the object's mass, v is its speed and r its distance from the sun at at its closest approach ("perihelion"). The third term of (G) is transformed thus:

$$\frac{m}{h^2} \rightarrow \frac{km/c^2}{(L/m_pc)^2} = \frac{kmm_p^2}{L^2}.$$

With these substitutions, equation (G) becomes

$$\frac{d^2u}{d\phi^2} + u = \frac{kmm_p^2}{L^2}, \tag{I}$$

where m is the mass of the sun and m_p is the mass of the orbiting object. Comparing the right-hand sides of the two equations (H) and (I) it appears that to a first approximation

$$\frac{kmm_p^2}{L^2} \doteq \frac{km}{c^2h^2} \qquad \text{or} \qquad h \doteq \frac{L}{m_pc} = r(v/c).$$

Look at the right-hand side of the geodesic equation (H). The ratio of the second term to the first term,

$$\left(\frac{3kmu^2}{c^2}\right) / \left(\frac{km}{c^2h^2}\right) = 3h^2u^2,$$

contains a factor h^2, so the second term is smaller by a factor of $(v/c)^2$ in comparison with the first term. This suggests that we try to solve the equation (H) with a solution

$$u = u_N + u_1,$$

where u_N is the Newtonian solution,

$$u_N = \frac{km}{c^2 h^2} \left(1 + e\cos(\phi - \phi_o)\right), \qquad \text{(J)}$$

and u_1 is expected to be a small correction. The constant e is the *eccentricity* of the elliptical orbit. For a perfectly circular orbit, $e = 0$; for Earth, $e \doteq \frac{1}{60}$; for Mercury, $e \doteq \frac{1}{5}$. The variable ϕ_o is the value of the angle when $t = 0$. Typically we set $\phi_o = 0$, but it does no harm to keep it in. Plugging $u = u_N + u_1$ into the equation (H) and again setting $k = 1$, $c = 1$ leads to

$$\frac{d^2 u_1}{d\phi^2} + u_1 = 3mu^2,$$

all the other terms canceling. Because u_1 is expected to be much smaller than u_N, for the second approximation we substitute $u = u_N$ into the term on the right, and obtain

$$\frac{d^2 u_1}{d\phi^2} + u_1 = \frac{3m^3}{h^4} \left(1 + e\cos(\phi - \phi_o)\right)^2$$

$$\doteq \frac{3m^3}{h^4} \left(1 + 2e\cos(\phi - \phi_o)\right)$$

because the square of the second term $e\cos(\phi - \phi_o)$ is much less than 1. Ignoring (temporarily) the 1 inside the parentheses (which will be picked up by u_N), the above equation has the form

$$\frac{d^2 u_1}{d\phi^2} + u_1 = A\cos(\phi - \phi_o)$$

(with $A = 6em^3/h^4$). One of its solutions is given by

$$u_1 = \tfrac{1}{2} A\phi\sin(\phi - \phi_o),$$

so that the solution to the second approximation of the geodesic equation is

$$u \doteq \frac{m}{h^2} \left(1 + e\cos(\phi - \phi_o) + \frac{3m^2}{h^2} e\,\phi\sin(\phi - \phi_o)\right).$$

This solution differs from the usual Newtonian solution by the term involving $\phi\sin(\phi - \phi_o)$. The coefficient of this term is proportional to $(m/h)^2$ (which turns out to be proportional to $(v/c)^2$). What does this do to the usual ellipse? Look at the Taylor expansion for cosine; if $y \ll x$ then

$$\cos(x - y) \doteq \cos(x) + y\sin(x),$$

so that

$$\cos(\phi - \phi_o) + \frac{3m^2}{h^2} \phi\sin(\phi - \phi_o) \doteq \cos(\phi - \phi_o - \Delta\phi),$$

where

$$\Delta\phi = \frac{3m^2}{h^2}\,\phi.$$

In one rotation, the angle ϕ increases by 2π radians. The extra angular distance the closest point (the perihelion) moves in one orbit is thus

$$\Delta\phi_{\text{orbit}} = \frac{6\pi m^2}{h^2}.$$

To figure out what this number is, we have to know something about ellipses. The relevant quantities are the semimajor axis a, the semiminor axis b, the eccentricity $e = c/a$ where

$$c = \sqrt{a^2 - b^2},$$

and the *latus rectum* ℓ:

$$\ell = \frac{b^2}{a} = a(1 - e^2).$$

The polar equation for an ellipse (really, *any* conic, determined by the value of e) is given by

$$\frac{1}{r} = \frac{1}{\ell}\left(1 + e\cos(\phi - \phi_o)\right),$$

which is to be compared with the Newtonian solution above, u_N (recall $u = 1/r$):

$$u = \frac{m}{h^2}\left(1 + e\cos(\phi - \phi_o)\right),$$

so, evidently, $h^2 = m\ell = ma(1 - e^2)$. Plugging this into the expression for $\Delta\phi$ above gives

$$\Delta\phi_{\text{orbit}} = \frac{6\pi m}{a(1 - e^2)} = \frac{6\pi km/c^2}{a(1 - e^2)} \tag{K}$$

where we replace m by km/c^2 to get mass in kilograms, not meters (see p. 315). Use these values:

$$
\begin{array}{lll}
k = 6.672 \times 10^{-11} \text{ m}^3/\text{kg-s}^2 & \text{(gravitational constant)} \\
m = 1.99 \times 10^{30} \text{ kg} & \text{(sun's mass)} \\
c = 2.998 \times 10^8 \text{ m/s} & \text{(speed of light)} \\
a = 5.791 \times 10^{10} \text{ m} & \text{(Mercury's perihelion)} \\
e = 0.20563 & \text{(Mercury's eccentricity)}
\end{array}
$$

and obtain

$$\Delta\phi_{\text{orbit}} = 5.021 \times 10^{-7} \text{ radian.}$$

This number is usually expressed in terms of seconds of a degree per century. Mercury needs 88 days to orbit, and there are 365.25 days in a year, so Mercury takes 415 trips around the sun in a century;

$$\Delta\phi_{\text{century}} = 415 \times 5.021 \times 10^{-7} \text{ rad} = 2.084 \times 10^{-4} \text{ rad.}$$

Convert this to degrees by using $180° = \pi$ radian;

$$\Delta\phi_{\text{century}} = 2.084 \times 10^{-4} \times \frac{180°}{\pi} = 0.01194°.$$

Finally, there are 60′ to a degree, 60″ to a minute, and hence 3600″ to a degree. Mercury's perihelion shift per century is thus given by

$$\Delta\phi_{\text{century}} = 0.01194° \times 3600″/1° = 43″.$$

This number had been a mystery to astronomers for more than sixty years. Pais (op. cit., p. 253, note[8]) writes that when Einstein saw that his theory explained the 43″, he felt perhaps the strongest emotion of his life, and later wrote to a friend, "For a few days I was beside myself with joyous excitement."

Incidentally, the diagram on p. 273 was produced through the command

```
PolarPlot[1/(1 + 0.5*Cos[0.97x]), {x, 0, 5.6 Pi}]
```

given to the computer program *Mathematica*. A polar plot

$$r = \frac{1}{1 + 0.5\cos(0.97\,\theta)}$$

made with a graphing calculator should give the same picture.

p. 284
[25] Here are the details provided by Eddington's treatise (pp. 90–91; d'Inverno, pp. 199–201). The geodesic equation (C) for light is

$$\frac{d^2u}{d\phi^2} + u = 3mu^2. \tag{L}$$

Equation (L) is actually the same as (C), though it lacks the third term. From (A) above, $(1/h) \propto ds$, but for light, $ds = 0$, so the third term in (C) equals zero if it is to describe light. Solve (L) by starting with the "associated homogeneous equation,"

$$\frac{d^2u}{d\phi^2} + u = 0.$$

This is solved by

$$u_o = A\cos\phi + B\sin\phi,$$

but choose A, B by requiring $u = u_{\max} = 1/r_{\min}$ when $\phi = 0$, and $u \to 0$ as $\phi \to \pm\pi/2$. This makes $B = 0$ and $A = 1/R$, where R is the radius of the sun. (The diagram on p. 280 is a little misleading; the light is supposed to graze the sun as it passes by, so the curve should be tangent to the circle which represents the sun.) That is, the first approximation to the light geodesic is

$$u_o = \frac{\cos\phi}{R}.$$

Now try to find a second approximation. As with the Mercury geodesic, try

$$u = u_o + u_1$$

(where u_1 is supposed to be small). Plug this into (L) but use only u_o in the correction term:

$$\frac{d^2(u_o + u_1)}{d\phi^2} + (u_o + u_1) = 3mu_o^2,$$

341

which reduces to

$$\frac{d^2 u_1}{d\phi^2} + u_1 = \frac{3m}{R^2} \cos^2 \phi.$$

After the usual trial and error we find a solution to be

$$u_1 = \frac{m}{R^2} \left(2 - \cos^2 \phi\right) = \frac{m}{R^2} \left(\cos^2 \phi + 2\sin^2 \phi\right)$$

(using the identity $\sin^2 \phi + \cos^2 \phi = 1$). The full solution (to the second approximation) is then

$$u = \frac{1}{r} = \frac{\cos\phi}{R} + \frac{m}{R^2} \left(\cos^2 \phi + 2\sin^2 \phi\right).$$

If we substitute $x = r\cos\phi$, $y = r\sin\phi$, $r = \sqrt{x^2 + y^2}$, we obtain

$$\frac{1}{r} = \frac{x}{rR} + \frac{m}{R^2} \left(\frac{x^2}{r^2} + \frac{2y^2}{r^2}\right).$$

Multiply by rR and put x by itself on the right, to find, finally

$$R - \frac{m}{R} \left(\frac{x^2 + 2y^2}{r}\right) = x,$$

which is exactly the equation in the middle of p. 284.

The diagram on p. 280 was obtained from *Mathematica* with the command

```
ParametricPlot[{Cos[t]/(Cos[t] + 0.1*(1 + (Sin[t])^2)),
Sin[t]/(Cos[t] + 0.1*(1 + (Sin[t])^2))}, {t,-Pi/4, Pi/4}].
```

Let, as usual, $x = r\cos\phi$, $y = r\sin\phi$. The equation for $u = 1/r$ is given above, and after a little algebra we obtain

$$r = \frac{R}{\cos\phi + (m/R)(1 + \sin^2 \phi)}.$$

Set the scale of the graph by setting the radius R of the sun to be equal to 1. In Einstein units, the mass of the sun is 1.475 kilometers, and the radius of the sun is 695,500 kilometers (about 109 times Earth's), so $m/R = 2.12 \times 10^{-6}$. Clearly this would be invisible if graphed, so increase the size of m/R to something larger, say 0.1. Also, this graph has the light beam passing on the right of the sun; Prof. Lieber shows the light passing to the left.

p. 285
[26] Given the equation of a straight line

$$y = mx + b,$$

the slope m is the tangent of the angle α made by the line with the x-axis. The tangent of the angle between two lines may be found with the formula for the tangent of two quantities α_1 and α_2;

$$\tan(\alpha_1 \pm \alpha_2) = \frac{\tan\alpha_1 \pm \tan\alpha_2}{1 \mp \tan\alpha_1 \tan\alpha_2} = \frac{m_1 \pm m_2}{1 \mp m_1 m_2}.$$

342

In the present case, the roles of x and y have been switched, so the slopes of the lines are $\pm 2m/R$. The angle α is the *difference* $\alpha_1 - \alpha_2$ of the two angles (because the slope of the second line is negative);

$$\tan \alpha = \frac{2m/R - (-2m/R)}{1 + (2m/R)(-2m/R)} = \frac{4mR}{R^2 - 4m^2}.$$

As $\alpha \to 0$, the tangent and the sine must have the same value; in the present case, $m \ll R$, and so this expression must reduce to the same value as $\sin \alpha$ (given on p. 285), as $m \to 0$, and they do: each approaches the value $4m/R$.

p. 286

[27] To make the comparison with experiment, put the mass in kilograms as before; use the values given in note [23] for k, m, and c, and take $R = 6.955 \times 10^8$ m for the sun's radius:

$$\alpha = \frac{4km}{Rc^2} = 8.48 \times 10^{-6} \text{ rad} = 4.86 \times 10^{-4} \text{ degree} = 1.75''.$$

Prof. Lieber does not relate some important historical aspects of this test. As she states, treating light as a Newtonian particle leads to a deflection only half of the relativistic prediction. The German physicist Johann Soldner calculated this in 1801. Einstein also obtained this value in 1912, from a "rough draft" of general relativity. An expedition to Brazil in 1912 was rained out, and a later one to Russia in 1914 was prevented by the outbreak of the First World War; if either of these had obtained good data, Einstein's 1912 prediction would have been proven wrong. Happily his final version of gravity in 1915 predicted the correct value, well before the 1919 eclipse. Not everyone was delighted with this brilliant validation of general relativity. The German physicist P. Lenard, perhaps motivated by nationalism or anti-Semitism, championed the much earlier (and erroneous) prediction of Soldner's over Einstein's. It may be this episode that Prof. Lieber refers to on p. 289. (See Pais, op. cit., pp. 199–200; 303–304, note[8].) Soldner's calculation is outlined in note [28].

p. 287

[28] Equation (J) describes the Newtonian motion of an object near the sun. If light were a stream of material particles, each with mass m' and speed c, equation (J) would correctly predict the deflection of light by the sun. (In fact light does *not* have mass, so the correct Newtonian prediction is that gravity does not deflect light *at all*.) In the standard units, (J) reads (with $\phi_o = 0$ and $h = (v/c)R = R$)

$$\frac{1}{r} = \frac{km}{c^2 R^2} \left(1 - e \cos \phi \right) = \frac{p}{R} \left(1 - e \cos \phi \right), \qquad \text{(J')}$$

where p is defined to be equal to $km/Rc^2 = 2.12 \times 10^{-6}$. The eccentricity $e = 2(K_{\max}/|U_{\min}|) - 1$ where $K_{\max}$ is the largest value of an object's kinetic energy, and $U_{\min}$ is the smallest value of its potential energy. Because c is so large, an object traveling at c does not appreciably speed up in falling toward the sun: we can take $K_{\max} = \frac{1}{2} m' c^2$. The expression for potential energy is $U = -kmm'/r$ where r is the distance between the light and the sun. This achieves its

343

least value when r takes its smallest value, namely $r = R$, the radius of the sun. Then

$$e = 2 \left(\frac{\frac{1}{2}m'c^2}{kmm'/R} \right) - 1 = \left(\frac{Rc^2}{km} \right) - 1 = \frac{1}{p} - 1 = \frac{1-p}{p}.$$

Then

$$\frac{1}{r} = \frac{p}{R} \left(1 - \frac{1-p}{p} \cos\phi \right) = \frac{1}{R}(p - (1-p)\cos\phi). \qquad \text{(K)}$$

Now repeat the analysis used in the relativistic case: determine the asymptotes for the hyperbolas this equation describes. To do that, get an equation for x and y out of equation (K). Cross-multiply, and set $x = r\cos\phi$;
$$R = rp - (1-p)x.$$

Rewrite as $rp = R + (1-p)x$, and square both sides, recalling that $r^2 = x^2 + y^2$:

$$(x^2 + y^2)p^2 = R^2 - 2R(1-p)x + (1 - 2p + p^2)x^2.$$

The term p^2x^2 occurs on both sides, so subtract it:

$$p^2y^2 = R^2 - 2R(1-p)x + (1-2p)x^2. \qquad \text{(L)}$$

To determine the asymptotes, let $x, y \to \infty$, and equation (L) turns into

$$p^2y^2 \doteq (1-2p)x^2 \Rightarrow y \doteq \pm\frac{\sqrt{1-2p}}{p}x,$$

so that the asymptotes have the slopes $\pm\sqrt{1-2p}/p$. Following the argument in note [26] or p. 285, the tangent of the Newtonian deflection α_N is given by

$$\tan\alpha_N = \frac{\sqrt{1-2p}/p - (-\sqrt{1-2p}/p)}{1 + \left(\sqrt{1-2p}/p\right)^2} = \frac{2p\sqrt{1-2p}}{1 - 2p + p^2}. \qquad \text{(M)}$$

The numerator of this fraction, by note [7], is approximately equal to $2p \times (1-p) = 2p - p^2$. Due to the tiny size of p, p^2 is completely negligible in comparison to p, so the numerator is not very different from $2p$, and the denominator is not very different from 1. That is,

$$\tan\alpha_N \doteq 2p = \frac{2km}{Rc^2} = \frac{2m}{R} \quad \text{(in relativistic units)}, \qquad \text{(N)}$$

which is almost precisely *half* the Einstein value,

$$\tan\alpha_E = \frac{4mR}{R^2 - m^2} \doteq \frac{4m}{R}.$$

p. 293
[29] Prof. Lieber's description of the mechanism by which atoms emit characteristic colors of light was not clear, and so was rewritten. Here are the differences between what she wrote originally and what now appears. Between the lines **The atoms of each element** and **And so**, the original text read

344

The atoms of each element
vibrate with a certain
DEFINITE period of vibration,
characteristic of that substance,
and, in vibrating, cause
a disturbance in the medium around it,
this disturbance being
a light-wave of definite wave-length
corresponding to
the period of vibration,
thus giving rise to
a DEFINITE color
which is visible in
a DEFINITE position in the spectrum.

Immediately following the sentence are present at S, and ending at
equation (71) becomes on the following page, the original read

Now then,
according to Einstein,
since each atom has
a definite period of vibration,
it is a sort of natural clock
and should serve as
a measure for an "interval" ds.
Thus take ds to be
the interval between
the beginning and end of one vibration,
and dt the time this takes,
or the "period" of vibration;
then, using space coordinates
such that

$$dr = d\theta = d\phi = 0,$$

that is,
the coordinates of an observer
for whom the atom is vibratiing at
the origin of his space coordinates
(in other words,
an observer traveling with the atom),

The line (p. 294) following as judged by an observer was originally

traveling with the atom

The lines (p. 295) following according to the above reasoning up to
DIFFERENT wavelengths were originally

would have a
DIFFERENT period of vibration,
and hence would emit light of a

The line (p. 296) following should have a read originally

slightly LONGER period of vibration

345

The editors hope that Prof. Lieber would not have been angry with us for these substitutions.

p. 297

[30] The gravitational red shift is difficult to measure from the light of distant stars: the atoms producing the light are moving so rapidly that it is not easy to separate the usual Doppler shift (due to their motion) from the gravitational shift. Worse, it turns out that the gravitational red shift does not test *general* relativity, because it is in fact a consequence of *special* relativity and the Equivalence Principle. An exquisitely precise test of the gravitational red shift was carried out by R. V. Pound and G. A. Rebka, with help from E. M. Purcell, at Harvard in 1959; the looked-for (and observed) effect was 5×10^{-15} cycle per second! See Clifford M. Will, *Was Einstein Right?* in Further Reading. The original paper—R. V. Pound and G. A. Rebka, "Apparent Weight of Photons", *Phys. Rev. Lett.* 4 (1960) 337–341—is available online: http://prl.aps.org/50years/milestones.

p. 297

[31] Sirius, also known as the Dog Star, is the brightest star in the night sky. In 1844 the German astronomer F. Bessel determined that Sirius was a binary star system, consisting of a fairly average star, Sirius A, about twice the mass of our sun, and Sirius B, the "companion," a scarcely visible white dwarf, the corpse of an average star, its nuclear fuel exhausted, now glowing only with residual heat. Sirius B has nearly the same mass as our sun, but packed into a volume about the same as the earth's. These two factors, a large mass with a small radius, make the gravitational red shift easier in principle to measure; the γ factor for Sirius is about a hundred times larger than for the sun, and three hundred thousand times that of the earth. The density of a white dwarf is about a million times that of water; a baseball-sized chunk of white dwarf matter would have a weight of half a million pounds!

p. 311

[32] The top equation is called Ampère's Law, and the bottom is known as Faraday's Law; the forms given are applicable only to regions without electric charge or magnets. These are usually written like this:

$$\text{Ampère: } \nabla \times \mathbf{B} = \frac{1}{c} \frac{\partial \mathbf{E}}{\partial t} \qquad \text{Faraday: } \nabla \times \mathbf{E} = -\frac{1}{c} \frac{\partial \mathbf{B}}{\partial t}.$$

$\mathbf{E} = (X, Y, Z)$ and $\mathbf{B} = (L, M, N)$ are respectively the electric and magnetic field vectors. In fact there are *four* Maxwell equations, and before Maxwell repaired it, Ampère's Law did not look as it does today (the right hand side was zero). Vector multiplications denoted by $\times$ (and $\cdot$) are explained in Taylor, *Classical Mechanics*, op. cit., pp. 6–7 (note [23]). Prof. Lieber called $\mathbf{E}$ and $\mathbf{B}$ the electric and magnetic *force* vectors throughout (this nomenclature is found in Einstein's own writings), but modern terminology reserves "force" for a quantity measurable in newtons.

346

p. 312

[33] Prof. Lieber is proving the *quotient theorem* for tensors. See Lawden, op. cit., §32 (note [12]). Prof. Lieber suppressed the indices on the tensor $X^{\alpha\beta\dots}_{\gamma\delta\dots}$; these have been added for clarity.

p. 316

[34] The most famous use of dimensional analysis in physics may be Lord Rayleigh's explanation of why the sky is blue, in 1871. See Peter Pesic, *Sky in a Bottle* (Cambridge, MA: MIT Press, 2005) 227–228, notes 37 and 38, for a clear explanation of how Rayleigh used units to arrive at the law explaining the color of the sky.

FURTHER READING

In the appendix to *Relativity*, Prof. Lieber provided a list of books entitled "Some Interesting Reading." While a few of these have historical importance, the rest have become outdated in the six decades since her book was published. Here are the four historically significant works, followed by newer texts, which better serve the purposes of her original references.

Eddington, Arthur S. *The Mathematical Theory of Relativity*, 3rd ed. Cambridge: Cambridge Univ. Press, Cambridge, 1937.

> The book Prof. Lieber refers to most frequently.

Einstein, Albert. *The Meaning of Relativity*, 5th ed. Introduction by Brian Greene. Princeton: Princeton Univ. Press, 2004.

> Lectures given at Princeton University in May 1921 and revised periodically until December 1954.

——. *Relativity: The Special and General Theories*. Introduction by Nigel Calder. New York: Penguin, 2006.

> Einstein meant for this to be a book for "the celebrated man (or woman) in the street."

Michelson, A. A., and E. W. Morley. "On the Relative Motion of the Earth and the Luminiferous Ether". *Am. Jour. Science* XXXIV (1887) 333–345.

> Also online at the American Institute of Physics web site: http://www.aip.org/history/exhibits/gap/.

~   ~   ~

D'Inverno, Ray. *Introducing Einstein's Relativity*. Oxford: Oxford Univ. Press, 1992.

> College-level textbook, exceptionally well written, with nearly all calculations shown explicitly. Prof. d'Inverno cites the first edition of the book in your hands as his inspiration for becoming a relativist; it seems only fair to return the compliment!

Holton, G., and Stephen G. Brush, *Physics, the Human Adventure: From Copernicus to Einstein and Beyond*. New Brunswick, NJ: Rutgers Univ. Press, 2001.

> Holton and Brush are outstanding physicist-historians, authorities in the areas of relativity and thermodynamics, respectively. This is one of the few really excellent textbooks available in paperback, unusually rich in historical detail.

Isaacson, Walter. *Einstein: His Life and Universe.* New York: Simon & Schuster, 2007.

> There are at least twenty biographies of Einstein in print. This one, written by a former managing editor of *Time* magazine, is the first to appear since all of Einstein's papers have been published. Special care is given to Einstein's science, treated non-mathematically, and his rich personal life.

Sartori, L. *Understanding Relativity: A Simplified Approach to Einstein's Theories.* Berkeley: Univ. of California Press, 1996.

> Special relativity is treated in algebraic detail while general relativity is discussed with very little mathematics.

Taylor, Edwin F. and John A. Wheeler. *Spacetime Physics*, 2nd ed. New York: W. W. Norton, 1992.

> High-school level introduction to special relativity, universally admired. John Wheeler (1912–2008) was Richard Feynman's doctoral advisor, and a major contributor to the revival of general relativity in the 1960's and 1970's. With his former students Kip Thorne and Charles Misner, Prof. Wheeler wrote the bible of general relativity, the graduate level text *Gravitation* (W. H. Freeman, 1972.) He also coined the name *black hole* and famously explained general relativity in a single sentence: "Matter tells space how to curve, and space tells matter how to move."

———. *Exploring Black Holes.* New York: Benjamin Cummings, 2000.

> An elementary introduction to general relativity, and the sequel to *Spacetime Physics.*

Thorne, Kip S. *Black Holes and Time Warps: Einstein's Outrageous Legacy.* New York: W. W. Norton, 1995.

> Recent developments in general relativity, described by a leading expert, with a minimum of mathematics.

Will, Clifford M. *Was Einstein Right?* New York: Basic Books, 1993.

> Testing general relativity, in plain English.